P9-CCI-422

Assessing Intelligence

Racial and Ethnic Minority Psychology Series

Series Editor: Frederick T. L. Leong, *The Ohio State University*

This series of scholarly books is designed to advance theories, research, and practice in racial and ethnic minority psychology. The volumes published in this new series focus on the major racial and ethnic minority groups in the United States, including African Americans, Hispanic Americans, Asian Americans, and Native American Indians. The series features original materials that address the full spectrum of methodological, substantive, and theoretical areas related to racial and ethnic minority psychology and that are scholarly and grounded in solid research. It comprises volumes on cognitive, developmental, industrial/organizational, health psychology, personality, and social psychology. While the series does not include books covering the treatment and prevention of mental health problems, it does publish volumes devoted to stress, psychological adjustment, and psychopathology among racial and ethnic minority groups. The state-of-the-art volumes in the series will be of interest to both professionals and researchers in psychology. Depending on their specific focus, the books may be of greater interest to either academics or practitioners.

Editorial Board

Please address all correspondence to the Series Editor:

Frederick T. L. Leong
The Ohio State University
Department of Psychology
142 Townshend Hall
1885 Neil Avenue
Columbus, Ohio 43210-1222
Phone: (614) 292-8219
Fax: (614) 292-4537
E-mail: leong.10@osu.edu

Assessing Intelligence

Applying a Bio-Cultural Model

Eleanor Armour-Thomas
Sharon-ann Gopaul-McNicol

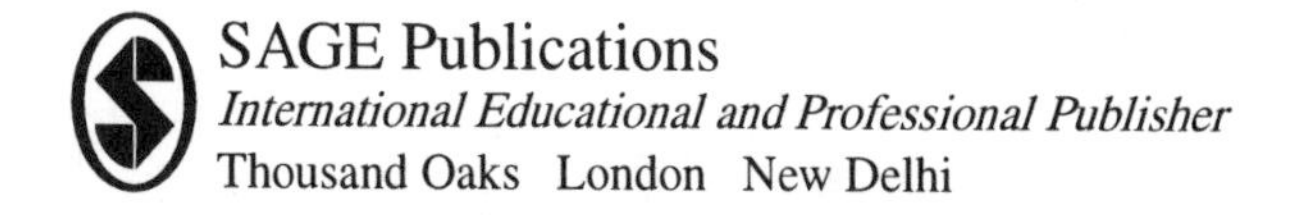

For information:

SAGE Publications, Inc.
2455 Teller Road
Thousand Oaks, California 91320
E-mail: order@sagepub.com

SAGE Publications Ltd.
6 Bonhill Street
London EC2A 4PU
United Kingdom

SAGE Publications India Pvt. Ltd.
M-32 Market
Greater Kailash I
New Delhi 110 048 India

Printed in the United States of America

Library of Congress Cataloging-in-Publication Data

Armour-Thomas, Eleanor.
Assessing intelligence: applying a bio-cultural model / by Eleanor Armour-Thomas, Sharon-ann Gopaul-McNicol.
p. cm. — (Ethnic minority psychology; vol. 1)
Includes bibliographical references and index.
ISBN 0-7619-0520-0 (alk. paper). — ISBN 0-7619-0521-9 (pbk.: alk. paper)
1. Intelligence tests. 2. Nature and nurture. 3. Cognition and culture. I. Gopaul-McNicol, Sharon-ann. II. Title. III. Series.
BF431.A578 1998
153.9′3—dc21 97-45335

This book is printed on acid-free paper.

98 99 00 01 02 03 10 9 8 7 6 5 4 3 2 1

Acquiring Editor: Jim Nageotte
Editorial Assistant: Fiona Lyon
Production Editor: Sherrise M. Purdum
Production Assistant: Lynn Miyata
Typesetter/Designer: Janelle LeMaster
Indexer: Juinee Uneide
Print Buyer: Anna Chin

To our families—*Bernard, Renaté, Bianca, Ulric, Monique Mandisa, and Monica*—with gratitude and love

And in loving memory of our parents
Cleaver and Celita Armour and *St. Elmo Gopaul*

Contents

Series Editor's Introduction

During the past two decades, Racial and Ethnic Minority Psychology has become an increasingly active and visible specialty area in psychology. Within the American Psychological Association, The Society for the Psychological Study of Racial and Ethnic Minority Issues was formed as Division 45 in 1986. This Division is now actively pursuing the publication of their own journal devoted to ethnic minority issues in psychology.

In APA we have also seen the publication of five bibliographies devoted to racial and ethnic minority groups. The first one was focused on Black males (Evans & Watfield, 1988) and the companion volume was focused on Black females (Iijima-Hall, Evans, & Selice, 1989). In 1990, the APA bibliography on Hispanics in the United States was published (Olmedo & Walker, 1990) followed by one on Asians in the United States (Leong & Witfield, 1992). The fifth bibliography was focused on North American Indians and Alaskan Natives and published in 1995 (Trimble & Bagwell, 1995).

As another indication of the increasing importance of racial and ethnic minority psychology, a brief review of the studies published on racial and ethnic differences in journals cataloged by Psychlit revealed a significant pattern. Between 1974 and 1990 (16 years), Psychlit cataloged 2,445 articles related to racial and ethnic differences. Between 1991 and 1997 (6 years), the number of articles related to racial and ethnic differences was 2,584. Put another way, between 1974-1990, an average of 153 articles were published each year on racial and ethnic differences. For 1990-1997, that number had jumped to an average of 430 articles published each year on racial and ethnic differences.

This pattern as well as many others have shown that racial and ethnic minority psychology is becoming an important and central theme in psychology in the United States. In recognition of this development, a new book series on **Racial and Ethnic Minority Psychology** (REMP) was launched at Sage Publications in 1995.

The REMP series of books is designed to advance our theories, research, and practice related to racial and ethnic minority psychology. It will focus on, but not be limited to, the major racial and ethnic minority groups in the United States (i.e., African Americans, Hispanic Americans, Asian Americans, and American Indians). For example, books concerning Asians and Asian Americans will also be considered. Books on racial and ethnic minorities in other countries will also be considered. The books in the series will contain original materials that address the full spectrum of methodological, substantive, and theoretical areas related to racial and ethnic minority psychology.

Within the field of racial and ethnic minority psychology, one of the earliest and recurring themes is concerned with the assessment of intelligence. This has been and remains a controversial topic in psychology as evidenced by the reactions to the work of Arthur Jensen and more recently to the publication of *The Bell Curve* by Herrnstein and Murray. The question of the so-called "significant racial differences in intelligence scores" remains a vexing one for psychology. From those who maintain that the differences in scores are real to those who argue that intelligence tests are biased against racial and ethnic minority groups, one thing is quite clear and stimulates little debate: Intelligence testing with racial and ethnic minority children and adolescents is not likely to disappear as a practice in the United States. It is therefore quite fitting that the first volume in the Sage series on **Racial and Ethnic Minority Psychology** is devoted to a well-reasoned and constructive approach to the use of intellectual assessment tools with racial and ethnic minority children and adolescents.

In this volume, Eleanor Armour-Thomas and Sharon-ann Gopaul-McNicol wonderfully integrate their many years of clinical experience in administering and interpreting intelligence test results on ethnic minority children to parents and teachers with the latest scientific knowledge. Based on the integration, they present a scientifically-based and culturally sensitive approach to using intelligence tests with racial and ethnic minority children and adolescents. Using a four-tier system, this integrated approach will be a significant contribution to the field and to the countless psychologists and educators who face the daily challenge of translating psychometrically-sound intelligence tests to the culturally complex and contextualized lives of their students, clients, families, and schools.

—Frederick T. L. Leong
Series Editor

Preface

In assessment circles, there seems to exist an implicit assumption that because teaching and learning concern the transfer and assimilation of knowledge and skill by persons uniquely equipped to do so, the assessment process should sample the pool of knowledge, skill, and competence. This logic seems to be based on the further assumption that if one can produce evidence of having mastered the assimilated knowledge and skill on demand, one not only knows but also can use these abilities whenever they are required. This basic conceptual model for assessment seems to ignore the fact that the traditional assessment process is also heavily dependent on the ability of the person being tested to recall and symbolically represent specific knowledge and to select iconic representations of skills. Some of us who have compared intellectual work in school and out of school have concluded that although these assumptions may be correct and may operate for some learners, there are vast differences between the ways in which mental work is experienced in school and in real-life settings. In real life, one actually engages in performances that contribute to the solution of real problems rather than producing, on demand and in artificial situations, symbolic samples of one's repertoire of developed abilities. In real life, one works with others to solve problems and often complements one's own knowledge and skill with those of others. Even more likely is the collective production of new knowledge and technique in response to experience with real problems that have special meaning to the persons encountering them. When we put these differences together with the relatively low correlations between scores on standardized tests and performance in real life, we recognize that there is some dissonance between what we typically do in the assessment of intellect and the ways in which humans exercise intellective functions in real life.

It is not only the arbitrary assumptions and processes that are associated with standardized tests that some consider to be problematic. In the latter part of this century, we have seen a plethora of activity directed at new approaches to educational assessment. Some of this ferment has been influenced by several years of agitation and complaint concerning the appropriateness of traditional standardized tests for the diverse populations served by modern schools. The most radical of these activities was the call, in the mid-1960s, for a moratorium on the use of standardized tests. Some courts placed restrictions on the use of such tests as the sole source of data in making certain decisions. At one point, some well-intentioned professionals called for the development of population-specific norms for groups known to show different patterns of scores on tests of educational achievement and intelligence. In recent years, we have seen calls for changes in the nature of what is measured, changes in the canons from which content is drawn, diversity in the performance modalities tapped by tests, the introduction of respondent choice, a shift to greater use of performance measures, and a variety of accommodations of test procedures to the special characteristics and needs of persons being tested. Considerable effort has recently been given to the design and implementation of performance assessment and portfolios. These adaptations and changes have been directed at the development of more "authentic" and valid measures of developed abilities and at making our tests more appropriate for diverse populations. Critics of standardized tests have argued that these proposed changes would improve the tests in addition to making them more "fair."

Perhaps it is the changing nature of the populations served by mass education and the changing criteria for what it means to be an educated person that have forced greater attention to this paradox. When the population was more homogeneous and society could absorb its school failures in a nonconceptually demanding workforce, the fact that schooling did not work for some of our members seemed less important. As the proportion of folks for whom school did not work increased, and as we became aware that even persons for whom schools once were adequate are not being enabled to function at intellective levels appropriate to changing societal demands, the potential crisis became more obvious. Teachers and schools became the targets of closer examination. Concern for how both of them can be held more accountable began to gain the focus of public attention. Because what teachers and schools produce is thought to be the achievements of their students, and because these achievements were increasingly viewed as inadequate, attention also came to be focused on educational tests and the processes by which we make judgments about the outcomes of schooling. It is this closer scrutiny that seems to have revealed to serious observers the contradiction between what it is we do in teaching and assessment, on the one hand, and optimally what should be happening in teaching, learning, and assessment on the other hand.

Despite this considerable amount of discussion and activity, it is interesting to note that the psychometric industry has not rushed to embrace these ideas. Rather, considerable effort has been directed at defending the validity of the extant psychometric technology and at seeking out and eliminating sources of bias in existing approaches to testing. The implicit assumption seems to be that the core practices are appropriate and do produce valid data, whereas many of the suggested changes present new problems, the most serious being the challenges of such innovations to the validity and reliability of extant assessment processes. It is a middle ground that is taken by the authors of this book.

It is in the context of this ferment for change and this assertion of stability and validity by the industry that Armour-Thomas and Gopaul-McNicol have produced this book. Rather than rejecting existing psychometric technology, these authors advocate its expansion to include a broader range of intellect-related behaviors and the contexts in which they are expressed. Asking the rhetorical question, "Are you a psychologist or a psychometrician?" they assert "that the psychologist is a clinician and a diagnostician, which means that he or she must go beyond IQ tests to assess intelligence, whereas the psychometrician relies purely on standardized tests of intelligence to determine a child's intellectual functioning" (p. 88). What is to form the basis for that clinical and diagnostic work and how the assessment is to be conducted constitute the focus of this book.

The central messages of this work involve (a) the origins and nature of intelligence, (b) a culture-embedded approach to assessment practices, and (c) the implications of a biocultural assessment system for policy determination and professional development. These three messages are delivered against a background of concern for the problems posed for assessment by the juxtaposition of increased concern with diversity in the population and broader respect for the pursuit of social justice in our society. The authors' treatment of issues related to practice, policy, and professional development isv grounded in the extensive discussion of the nature of intelligence with which they open this work.

Approximately a quarter of a century ago, Anne Anastasi suggested that we might better think of intelligence as the manifestation of developed mental abilities. She was trying to avoid the problems associated with those conceptions of intellect that privileged genetically determined and fixed states. Without addressing the nature-nurture controversy, Anastasi asserted that intellect is best thought of as a product—a result. No matter what the source, what is being measured is what has been developed. From those developed abilities, we infer the quality or level of intelligence.

In this book, we find a similar view of intelligence. The authors assert some biological substrate, but they think that the quality of cognitive function reflects the organism's interactions with its physical and cultural ecology. Consequently, to understand the character and quality of intellective development, its assessment must include these broader dimensions of cultural and ecological diversity, in which and as a result of which developed abilities are shaped. The authors argue that three interrelated and dynamic dimensions of intelligence can be identified. They are (a) biological cognitive processes, (b) culturally coded experiences, and (c) cultural contexts. They assert that these interacting multidimensional aspects of intellect must also be respected in the assessment process. To capture the complexities of this theoretical model, a four-tier biocultural assessment system is proposed that includes (a) psychometric achievement, (b) psychometric potential, (c) ecological conditions, and (d) other intelligences. In Part II, the authors describe the system and its use in the context of a critique of standardized tests. Part III is devoted to discussions of implications for professional development and public policy. We are indebted to the authors for a thoughtful analysis of a complex set of issues and for a rational approach to psychometric practice that is informed by their biocultural model for understanding human intellective competence.

—Edmund W. Gordon
John M. Musser Professor
of Psychology, Emeritus
Yale University

Acknowledgments

To our immediate family—Bernard, Renaté, Bianca, Monica Gopaul, Ulric, and Monique Mandisa—we extend our most profound gratitude. Your love, patience, and unflagging support throughout this challenging time will be remembered in our hearts forever.

To our parents, who instilled in us a sense of confidence and pride about "being the best you can be," we extend much gratitude, respect, and tenderness.

To our siblings, Cleavy, Edward, Margaret, Gail, Wendy, Kurt, and Nigel, our nieces and nephews, Ricardo, Brian, Jerry, Paula, Gena, Catrice, Wesley, Reid, Michael, Kja, Michelle, Skyler, and Kris, and our cousins, Doreen, and Maria, we thank you for your ongoing support and patience.

To our professional colleagues and special friends who have helped to nurture a respect for cross-cultural inquiry, we thank you immensely. In particular, we extend our appreciation to the following individuals: Norris Haynes, Miriam Azaunce, Coleen Clay, Brenda Allen, Ernest Washington, Randolf Tobias, David Rollock, Jesse Vasquez, Earl George, Janet Brice-Baker, Ann Francis-Okongwu, Nancy Boyd-Franklin, A. J. Franklin, Frederick Harper, Wade Boykin, Veronica Thomas, James Williams, Mary Conoley, Constance Ellison, Nicola Beckles, William Proefriedt, Aldrena Mabry, Lauraine Casella, Arthur Dozier, George Irish, Delroy Louden, Michael Barnes, Dawn Arno, Alice Artzt, Emery Francois, Jefferson Fish, Alina Camacho-Gingerich, Willard Gingerich, Tony Bonaparte, Frank Leveness, Tony Gabb, Erika Wick, James Curley, Joya Gomez, Charmaine Edwards, Headley Wilson, Susan Lokai, Seretse McHardy, Koreen Seabrun, Sandra Hosein, Jennifer D'Ade, and Joanne Julien.

To our research assistants—Grace Reid, Orlean Brown, Steve Choi, Renaté and Bianca Thomas, Samuel Korobkin, Andrew Livanis, Sibel Diaz, Robert Schmidt, Jessica Nava—and other students in training who provided much support, we extend our gratitude.

To the leading thinkers in the field who most influenced our work—Edmund W. Gordon, Lev Vygotsky, Janet Helms, Robert Sternberg, Ulric Neisser, Stephen Ceci, Barbara Rogoff, Howard Gardner, Jean Lave, Michael Cole, Stanley Sue, Jim Cummins, Robert Williams, A. Wade Boykin, Asa Hilliard, III, Esteban Olmedo, and Richard Figueroa—we extend our deepest respect and gratitude.

To the communities and students of Queens College, Howard University, and St. John's University, we thank you for your support in the nurturing of our thinking.

To our deceased loved ones, Cleaver and Celita Armour, St. Elmo Gopaul, Christopher Edwards, and Josephine Fritzgerald, may God bless you all.

Introduction

This book is divided into three parts. Part I examines the issues pertaining to intellectual assessment in a multicultural society (Chapter 1). The conceptions of the nature of intelligence are discussed in Chapter 2. Part I also delves into biological influences on cognition as well as cultural influences on cognition (Chapter 3). Chapter 4 proposes a new perspective of intelligence—a biocultural model developed by the authors.

Part II explores a new model for assessing all children—the four-tier biocultural assessment system (Chapters 5 and 6). Chapter 7 provides a critical review of four commonly used intelligence tests—the Wechsler scales, the Stanford Binet, the Kaufman, and the Woodcock. Chapter 8 presents sample reports, culminating in the most culturally sensitive psychological report based on the four-tier biocultural assessment system. Finally, an evaluation of the assessment system is performed in Chapter 9.

Part III offers training suggestions for teachers, parents, counselors, and psychologists for enhancing the intellectual potential of all children (Chapter 10). Finally, implications for future research and clinical work, as well as a vision for policymakers as to how to ensure culturally sensitive assessment and tutelage, are suggested in Chapter 11.

PART I

Intelligence: Major Issues and Challenges

1

Intellectual Assessment in a Multicultural Society

Unlike other reform movements in U.S. education, educational goals in the 1990s are centered around the three E's: economics, excellence, and equity. The national prosperity of the nation is perceived as dependent on the capacity of schools to produce workers with deep conceptual understanding in various domains of knowledge and demonstrations of advanced skills of reasoning, problem solving, and higher-order thinking. Although concern for equality of educational opportunity is not a new concept in U.S. education, it is new to envision academic excellence for all children, irrespective of race, class, ethnicity, gender, language, or any other dimension of human diversity by which individuals and groups are categorized in the society. This vision of education for the nation's children seems incompatible with the continuing use of intelligence test data to make selection and placement decisions of children for certain types of educational program. Implicit in the use of standardized intelligence test data for these purposes is a notion that differences in test scores reflect fundamental differences in intellectual capacities and that the goal of assessment is to objectively sort children according to these underlying differences. Once accurately sorted, educational programs can be more readily matched to these underlying differences. The scientific legitimacy of this notion, however, is in serious question when judgments about intellectual capacity and educational placements parallel race, class, and language differences. For example, it is a well-documented finding that the average performance of children from African descent differ by as much as one standard deviation on any standardized test of intelligence. It is also common knowledge that racial and ethnic minorities judged as "low ability" are placed in low-tracked and remedial classes or in special education programs (Persell, 1977; Rosenbaum, 1980; Slavin, 1987). What exactly do differences in IQ scores mean and what are the origins of such variation have been controversial questions since Jensen's (1969) declaration that the differences are real and substantial—a sentiment echoed in Herrnstein and Murray's book, *The Bell Curve* (1994).

In this chapter, we examine the principles of equal educational opportunity and multiculturalism and show how the purposes and practice of intelligence testing are at odds with those principles. The chapter begins with a discussion of key concepts in educational opportunity that serves as an entry point to a critical examination of the selection and placement purposes of standardized tests of intelligence. Next, a discussion of key issues in multiculturalism follows that also serves as an entry point for a critical examination of the assumptions of standardized tests of intelligence as these relate to children from certain racial and ethnic groups in U.S. society. The chapter ends with a brief exploration of new directions for intellectual assessment for this population of children.

Equal Educational Opportunity

The concept of equal educational opportunity reflects a core principle of democracy that ensures that all children, regardless of race, ethnicity, culture, language, gender, and other socially constructed characteristics of persons, should be entitled to equal access to the best and most appropriate education available. Gordon and colleagues (Gordon & Bonilla-Bowman, 1994; Gordon & Shipman, 1979) distinguish this notion of distributive equality from the concept of distributional appropriateness and sufficiency as essential criteria for educational equity. Building on Rawls's (1973) principles of justice, Gordon and Bonilla-Bowman (1994) argue that

> Educational treatments like medical treatments must be appropriate to the condition and characteristics of the person being treated and sufficient to their support and correction. To give all patients sulfa drugs when some need penicillin does not meet the condition of appropriateness. To give all one dose when some need three doses does not meet the condition of sufficiency. (p. 30)

Thus, Gordon introduces social justice attributes of need and circumstance criteria into the characterization of educational equity such that the emphasis is not on providing all children with the same educational treatments but rather on the appropriateness and sufficiency of treatment to the functional needs and characteristics of the persons being educated.

In a somewhat different vein, Howe (1992) calls attention to three key issues that any defensible interpretation of the principle of equality of educational opportunity must accommodate: (a) freedom and opportunities worth wanting—that is, the ability to deliberate effectively, and the opportunity to exercise it. This latter condition requires the availability of necessary information for deliberation and that social conditions do not serve as constraints or barriers for persons acting on the results of deliberations that are disproportionate to the burden of other deliberators; (b) equal educational opportunity as enabling—that is, taking advantage of early educational opportunities enables the acquisition of other ones later that, in turn, serve as the gateway for other societal goods such as adequate income, desirable employment, and political power; and (c) equal opportunity and children—that is, until they attain a certain age, children are not likely to pursue freedom and opportunities genuinely worth wanting or enabling. Schools, parents, or both, however, should ensure that one day children should be able to enjoy these benefits.

Both Gordon's and Howe's conceptions of equal educational opportunity make no conditional constraints on access to the nation's resources for all its children, regardless of the amount or quality of their intellectual endowments and the source of its distribution. Historically, variation in performance on standardized tests of intelligence has been used as a basis to select and place children in programs that are worth wanting or enabling for some children. For other children, however, such programs are neither worth wanting or enabling.

Consider the following findings of a *U.S. News & World Report* ("Separate and Unequal," 1993, p. 54) analysis of data from the Department of Education in 39 states:

Retarded: Black, 26%; white, 11%; Hispanic, 18%
Learning disabled: Black, 43%; white, 51%; Hispanic, 55%
Emotionally disturbed: Black, 8%; white, 8%; Hispanic, 4%
Speech impaired: Black, 23%; white, 30%; Hispanic, 23%

Compared to their percentage in the overall student population, black and Hispanic students are overrepresented in some special education programs. Closer examination of these statistics shows that in comparison to whites, a higher proportion of blacks are labeled mentally retarded and emotionally disturbed. They are less likely than whites, however, to be labeled with less stigmatizing labels of speech impaired or learning disabled.

A deficit explanation of biological inferiority is likely to be the heredity view for the overrepresentation of racial and ethnic minorities in special education programs. An alternative explanation and one that we support, however, has to do with a complex mix of factors involving (a) teaching and learning experiences reflective of the dominant group values in U.S. society, (b) the learning experiences of children reflective of the sociocultural milieu in which they are socialized, (c) a categorical approach to intellectual assessment that locates the problem of academic and intellectual dysfunction within the students themselves, and (d) the use of IQ tests that reflect a value system and experiences similar to those of the dominant group in the society and the schools that serve their children. We contend that many racial and ethnic minority children, particularly those from a low-income background, experience academic difficulty in adjusting to the norms and expectations of the dominant group values in U.S. society as reflected in the curriculum, instruction, and assessment practices in the classroom. Without appropriate guidance and support or an incentive system that works for them, these children experience academic difficulties so acute that inevitably they are referred for a psychological evaluation that includes an achievement test and a conventional IQ measure. Because IQ and academic achievement tests are correlated, it is not surprising that low to very low scores are obtained on both measures, thereby making plausible, although erroneous, the location for the source of the academic difficulties within the children themselves. The dominant categorical model so pervasive in intellectual assessment and special education will most likely identify a category among its complex array of labels (mentally retarded, mildly retarded, slow learner, specific learning disability, emotionally disturbed, language impaired, etc.) to match the "observed" dysfunction. This explanation is consistent with those of Cummins (1984, 1991), Mercer (1979), and Samuda (1975).

Other critics question whether standardized test data should even be used to inform educational decision making and instruction. For example, Glaser (1977) claimed that such tests give go/no-go selective decisions but do not provide information that is sufficiently diagnostic for the conduct of instruction. Gordon (1977) made a similar point when he argued that at best, test data analysis provides gross characterization of success and failure of students in relation to some reference group, but such analysis provides little information about the process by which individuals engage the task and is insensitive to an individual's differential response tendencies. As Gordon asserted, however, the description of these characteristics of behavioral individuality is of crucial importance in the design and management of teaching and learning transactions.

The Association of Black Psychologists (Williams, 1970) summed up the disillusionment of critics of the use of mental test data for children of African heritage when it charged that tests label black children as uneducable, place black children in special classes, potentiate inferior education, assign black children to lower-education tracks than whites, deny black children higher-educational opportunities, and destroy positive intellectual growth and development of black children (p. 5). Almost 30 years later, the skepticism about the validity of intelligence tests for making educational decisions for racial and ethnic minorities still remains in many quarters (Armour-Thomas, 1992; Cummins, 1984; Figueroa, 1990; Gopaul-McNicol, 1992a; Hilliard, 1991; Lipsky & Gartner, 1996).

The specter of biological determinism and inequitable educational opportunity pose a serious moral dilemma for a society committed, at least in principle, to the ideals of equal educational opportunity for all its children irrespective of the cultural and linguistic background from which they come. We turn next to a discussion of the principle of multiculturalism and consider what it means for the intellectual assessment of children whose cultures are different from the culture of the dominant group in our society.

Multiculturalism

Multiculturalism is the term used to describe the existence of groups within a society that share ways of life or cultures that are rooted in distinct histories. Defining attributes of these ways of life include norms, values, and beliefs that govern the daily interactions of its members, that give meanings to these interactions, and that help shape the development and maintenance of a common identity. This perspective of cultures is consistent with those of the noted anthropologist Geertz (1973), who conceived of these sets of shared meanings as webs of significance that provide organization and maintain the cohesiveness of daily life of members of a cultural group.

Multiculturalism described in this way suggests to us that different cultures would have different perspectives that embody different value systems about how its members should behave among themselves as well as toward others outside their group. There is more, however, than simple recognition that different value systems exist in a multicultural society such as the United States. There are expectations for how different cultural groups should engage each other in conversations about their respective diversities. There must be not only acknowledgment of members of a cultural group's moral right to define their "webs of significance" on their terms but also a responsibility that other cultural groups will respect their right to do so. These norms of engagement are not easy to implement, as Moon (1993) pointed out:

> Moral pluralism does not simply involve the existence of . . . incompatible values—values that may come into conflict with each other depending upon circumstances. Rather, it arises when different people resolve these conflicts in systematically different ways, or when they come to hold ends or principles that are inherently incompatible. (p. 23)

Of course, there are a number of ways that democratic societies such as the United States seek to accommodate multiculturalism (e.g., perspectives of tolerance, common values, and search for universals). When we get beneath the surface rhetoric of these seemingly enlightened positions, however, the terms of accommodation or standards of acceptability are neither apolitical or morally neutral. Indeed, in a provocative essay on how dominance is concealed through diversity, Boyd (1996) pointed out the inadequacies of current perspectives on cultural pluralism in dealing with the dilemma of diversity: accommodating cultural relativism within a broader evaluative commitment to some generalizable morally binding constraints on all individuals and groups. Too often, according to Boyd, failure to deal with the dilemma leaves the prescriptive preferences of the dominant view in control.

Ostensibly, the idea of dominance in any form is incompatible with the principle of equity, a central tenet of U.S. democracy. It seems almost heretical to consider the notion of asymmetrical power relations among groups in a society for whom the metaphor *e pluribus unum* holds such significance in its developmental history. The seemingly intractable nature of the problem of access to education, however, essentially and sufficiently responsive to the needs of some learners, has led many to question whether the principle of equal educational opportunity operates in the same way for all children. The disproportionate and persistent nature of the problem of underachievement of some racial, linguistic, and ethnic groups and the centrality of the use of IQ tests in both explaining the reason for the underachievement and making subsequent educational placements have fueled the suspicion that something else is in the equity pot!

Over the years, scholars from diverse disciplines have given name to the "something else" and have speculated with respect to how it functions. For example, DeVos (1984) used the term *caste thinking* to characterize the belief that there is some biological, religious, or social inferiority—unalterable—that distinguish members of a minority group from those of the dominant group in the society. He thinks that dominant-group members usually rationalize their exploitation of others with this form of thinking. Using this concept, Ogbu (1987) coined the term *castelike minorities* to refer to those groups that have been incorporated into the United States through conquest (e.g., Native Americans and Mexicans) or through slavery (individuals of African ancestry). Ogbu, like DeVos, believes that these historically subordinated groups have been the object of systematic discrimination in all realms of experience, including the provision of inequitable educational experiences. Other scholars use the terms *hegemony* (Apple, 1979; Boykin, 1983), *communicentric bias* (Gordon, Miller, & Rollock, 1990), and *eurocentrism* (Asante, 1988) to describe essentially the same phenomenon and to call attention to the inability of some members in the society who are too vulnerable or powerless to escape its pernicious effects.

Considerations of these issues of multiculturalism led us to suspect that a hidden dominance lurks within the structure and practice of standardized tests of intelligence. More specifically, we think that children from certain racial and ethnic minority groups (e.g., Native Americans, African Americans, Mexican Americans, and Caribbean Americans) are not allowed the right to develop and express their intelligence on their terms. Rather, their intellect is constantly being subjected to comparisons with other cultural groups (children from European descent) using criteria that "prove" their intellectual incompetence over and over again. In the next section, we elaborate on this claim of eurocentric bias through a critical analysis of the underlying assumptions of standardized tests of intelligence. In the process of this analysis, we examine how these assumptions are problematic when applied to children whose culture is substantively different from that of the dominant group in U.S. society.

Assumptions of Standardized Tests of Intelligence

Generally, proponents of standardized measures of intelligence justify the comparison made about intellectual performance between different cultural groups on the grounds that such measures meet at least four implicit criteria: (a) The tasks as constructed are culturally fair—that is, items do not favor any particular cultural group; (b) the tasks assess the cognitive abilities underlying intellectual behavior; (c) the tasks could sufficiently elicit the deployment of particular mental operations; and (d) accurate interpretations could be made from comparing the average IQ scores of different cultural groups. When these criteria are applied in the assessment of intellectual functioning of children from certain racial and ethnic groups, numerous difficulties arise that raise serious questions about the validity of comparative judgments about their performance. In the following sections, some problems of standardized tests for these populations are discussed.

Item Selection Bias

Any intelligence test construction procedure involves the standardization of the test on a representative sample. This means, in the case of U.S. society, the majority of subjects will come from the dominant group—Anglo- or European Americans. A minority of groups will make up the rest of the sample—groups of non-European origin including Native Americans, African Americans, Mexican Americans, Caribbean Americans, Asian Americans, and Pacific Islander Americans. During the early stage of item development, the majority of items selected for tryout will obviously reflect the prior learning experiences of the Anglo- or European American group. Because academic achievement correlates with any standardized test of intelligence, this means that the learning experiences are similar to those acquired through schooling.

It cannot be assumed that the learning experiences of the majority group are similar to those of the minority groups (e.g., Boykin, 1986; Ogbu, 1986). Nor can it be assumed that the learning experiences derived from schooling for majority children are similar to those of minority groups, particularly those from low-income backgrounds (e.g., Oakes, 1990). Even if items reflective of the prior learning experiences of minority groups were included in the early phase of item development, they would more than likely be screened out during item analysis. A common psychometric practice in test construction is to retain only those items that correlate with the total test and that are of moderate difficulty. Items that are appropriate for minority groups more often than not will be difficult for the majority group and will not correlate well with the total test (e.g., Williams's [1975] Black Intelligence Test of Cultural Homogeneity). Thus, the process of item selection for an intelligence test does involve some bias against minority groups (see Cummins, 1984, for a comprehensive discussion of the issues pertaining to item selection bias).

Lack of Specificity About Cognitive Processes

Although psychometric studies of cognitive abilities have identified a number of cognitive processes that underlie performance on intellectual tasks (e.g., Carroll, 1993; Horn, 1991a), there is no consensus about how many and which processes combine to produce behavior indicative of intelligence. Horn (1991b), in describing the difficulty of identifying these processes with precision, remarked that "Specifying different features of cognition is like slicing smoke—dividing a continuous, homogeneous, irregular mass of gray into . . . what?" (p. 198).

Lack of specificity of cognitive processes when describing intellectual behavior in assessment situations is likely to favor some cultural groups but hinder others. Almost three decades ago, Messick and Anderson (1970) called attention to the fact that these same tests may measure different processes in minority children from low-income backgrounds than in white middle-class children. A similar concern was articulated by Farnham-Diggory (1970), who suggested that the multiplicity of cognitive processes in Thurstone's Primary Abilities

Test made it difficult to determine which of these processes pose difficulties for children of African heritage. In our own work, it was not always clear that, when children from culturally and linguistically diverse backgrounds answered items incorrectly, these errors were due to inaccurate or inefficient deployment of cognitive processes.

Content Bias

When comparisons are made between cultural groups on standardized measures of intelligence, inaccurate assumptions about aspects of the task requiring the deployment of mental processes may be made. In other words, differences in cognitive processes may be a function of variability in dimensions of tasks within a particular cultural context. Many cross-cultural studies have reported that the familiarity or unfamiliarity of content may constrain or promote the efficient or accurate elicitation of cognitive processes in tasks involving memory, reasoning, perception, and problem solving (e.g., Cole, Sharp, & Lave, 1976; Greenfield, 1974; Laboratory of Comparative Human Cognition [LCHC], 1982; Lave, 1977; Rogoff, 1981; Saxe, 1988). In these studies, researchers focused on the domain-specific thinking of people as they solve problems routinely encountered in their culture. For example, in an early cross-cultural study, Gay and Cole (1967) assessed classification competencies between schooled and unschooled Liberians and U.S.-schooled children by using tasks involving culturally appropriate content: bowls of rice and geometric blocks. African subjects performed as competently when the tasks involved rice as the stimulus as their U.S. counterparts who used geometric shapes. In contrast, decrements in performance were observed for both groups when the contents of the stimulus materials were reversed. In other words, African children classified geometric shapes as poorly as the U.S. children sorted rice. Recently, several researchers have used concepts such as everyday cognition (e.g., Guberman & Greenfield, 1991; Rogoff & Chavajay, 1995) and practical intelligence (Neisser, 1979; Sternberg, 1985; Sternberg & Wagner, 1986) to better understand the thinking competencies of individuals and groups that underlie their performance on culturally familiar tasks in their own communities.

The LCHC (1982) used the concept of "functional stimulus equivalence" to account for these content-specific findings. This means that to make valid comparisons of intellectual performance between cultural groups, it is critical that the stimulus attributes of the task be equivalent for both groups in their ability to elicit the cognitive processes under investigation. From our experiences with standardized tests of intelligence with children from culturally diverse backgrounds, it does not appear that test designers have sufficiently addressed the issue of functional equivalence in item construction or selection.

Unfamiliar Sociolinguistics

One of the hallmarks of intelligence testing is the adherence to standardized procedures during its administration. One aspect of these procedures has to do with examiner-examinee interactions. When to probe for more information, how to respond to the examinee's queries, and how much feedback to give are all governed by a strict protocol using a standard format. Such constraints of the testing environment, however, may preclude an accurate estimation of intellectual competence of some children from culturally diverse backgrounds. An example of this problem concerns the dynamics of sociolinguistics among ethnic minority children from low-income backgrounds. Sociolinguistic variables are courtesies that govern verbal interactions that can have a positive or negative impact on student motivation to engage in a task. Miller-Jones (1989) provided some excellent examples of the verbal interactions between an examiner and an African American child that demonstrate how such interactions of the testing context may result in erroneous judgments of intellectual functioning of that child. The question and response exchange clearly indicated that the child did not understand or was not familiar with the social dynamics within the testing context. In short, in a standardized testing context, the norms of discourse are predetermined, and performance differences could result to the extent that children's sociolinguistic patterns of communication are different from the examiner's script.

Test Interpretation Bias

Comparisons of intelligence test scores on the basis of ethnic classification are likely to lead to bias in interpretation because there is considerable heterogeneity within such groups in terms of family structure, geographical region, social class, language, and education (e.g., Harrison, Wilson, Pine, Chan, & Buriel, 1990; McLoyd, 1990). For example, Phinney (1996) identified at least the following three aspects of ethnicity for ethnic groups of color in the United States (e.g., Native Americans, African Americans, Latinos, Asians, and Pacific Islanders) that may have psychological relevance for influencing behavior:

- Cultural values, attitudes, and behaviors that differentiate ethnic or racial groups
- Identity—the perceptions of what it means to the individual to belong to an ethnic or racial group
- Experiences associated with minority status in the United States (e.g., discrimination, prejudice, and a sense of powerlessness)

These features of psychological characteristics are similar to those identified by Sue (1991), who, like Phinney (1996), viewed them as overlapping and confounded constructs.

Thus, an ethnic label alone is not enough to interpret differences in performance between ethnic groups. To understand intellectual behavior would require an unpackaging of the multiple psychological processes associated with the label as a number of researchers have recommended (e.g., Betancourt & Lopez, 1993; Phinney, 1996; Poortinga, van de Vijver, Joe, & van de Koppel, 1989; Whiting, 1976).

One of the methodological difficulties associated with the unpackaging of psychological processes, however, is that researchers and pollsters may not have fully understood or appreciated their importance in influencing behavior. As Betancourt and Lopez (1993) noted, research instruments and surveys often require subjects to indicate race by selecting among confounded status variables (e.g., white, black, Latino, Native American, and Asian). Zuckerman (1990) cautioned that the loose way in which race, culture, and ethnicity are treated may contribute to interpretations of findings of observed differences between groups that reinforce racist conceptions of human behavior.

Even in cases in which researchers try to minimize the confounding of status characteristics, it is not clear that they succeed. For example, by statistically controlling for the effects of socioeconomic status among Latinos or Asians children in a study of, for example, intelligent behavior, a researcher can unwittingly mask the effects of language and culture. To the extent that these two latter characteristics are differentially experienced by members who are located in different class positions, one can misattribute to socioeconomic status the influence of language or culture or both.

Comparisons of average IQ test scores of black and white children are equally problematic when using the race label to make group classifications. As Helms (1992, p. 1085) noted, tremendous variation may exist within black and white populations because of

a. Voluntary and involuntary interracial procreation
b. The tendency of researchers to assign subjects to one group or another on the basis of physical appearance
c. The decision of some visible racial or ethnic groups that appear white to disappear into white society (a process called passing in black culture)
d. The possibility that immigrants who would be considered black if they were born of similar parentage in this country classify themselves as white or other than black

Variations in perceptions of race also exist in other countries. In a recent review of conceptions of race as a socially constructed phenomenon, Eberhardt and Randall (1997) described the fluid continuum by which race is defined in Brazil, Latin America, and Caribbean countries. Factors such as economic and geographic mobility are allowed along the continuum as are a wide range of colors. In Caribbean and Latin American countries, it is not uncommon for individuals to change color or to have different perceptions of physical attributes such as hair texture and skin pigmentation as a function of economic mobility. According to these researchers, in these countries, money "whitens."

Thus, an interpretation of differences of IQ scores between subjects of European and African decent on the basis of racial categorization is indefensible and downright racist.

New Directions for Intellectual Assessment

The discussion of equity and multiculturalism principles in relation to standardized tests of intelligence raises many troubling questions about their selection and placement purposes and its use with children from certain racial and ethnic groups in U.S. society. It is our view that in a multicultural society, nondiscriminatory assessment would require an informed understanding and appreciation of the cultural influences on children's intellectual behavior. Furthermore, it would also require an understanding of the mechanism by which such factors function to promote or constrain the deployment of cognitive processes. As we argue in Chapter 5, no one measure can fulfill these multiple expectations of intellectual assessment. Rather, multiple forms of assessment need to be developed that

- Sample a broad range of cognitive processes
- Sample content that is functionally equivalent for the groups targeted for assessment
- Are sufficiently diagnostic so as to uncover strengths and weaknesses of manifest cognitions as well as emerging cognitive potentials
- Allow sufficient time to demonstrate accurate use of cognitive processes
- Are sensitive to the sociolinguistic patterns that children bring to the assessment environment
- Provide a scaffold for easier elicitation of the cognitive processes in a given domain of knowledge
- Allow opportunity to learn the process and knowledge demand of the task
- Assess the manifestation of these processes in more real-world environments
- More precisely inform prescriptive pedagogical or rehabilitative interventions

A fuller development of these issues is found in Chapter 6, in which we lay out in detail the four-tier approach to assessing intelligence.

Conclusion

A democratic society committed to equity and multiculturalism requires that fair and nondiscriminatory measures be used to assess children's intellectual competencies regardless of their background characteristics. When judgments about intelligence parallel race, language, and ethnicity and when such judgments are used to make educational placements of dubious quality, the principle of equity is doubly compromised for children. We hope, however, that increased awareness and genuine respect for cultural pluralism will lead to the development and use of measures that have greater psychodiagnostic and prescriptive utility than those that currently exist. A just and humane society demands no less for its children.

2

Conceptions of Human Intelligence and Implications for Its Assessment

"What is human intelligence?" is one of the most intriguing questions in the study of psychology. As early as 2000 years ago in ancient China and Greece, philosophers pondered this question as have the psychologists in the 20th century, but a definitive response remains elusive. When, in 1921, the "Journal of Educational Psychology" published a number of papers on intelligence written by distinguished psychologists, there was remarkably little agreement among the 14 definitions ("Intelligence and Its Measurement," 1921). Recently (Sternberg & Detterman, 1986), a symposium was held to reexamine the concept but, yet again, the experts still do not agree. The theoretical ambiguity about the construct makes it difficult to gather valid and reliable evidence about intellectual behavior or to make accurate interpretations and inferences from differences in IQ scores. The absence of consensus does not mean inconsequential knowledge about the nature of the construct or its measurement because numerous handbooks and journal articles have been published on both topics during the past 90 years. Despite the continuing proliferation of perspectives of intelligence and its assessment, the field of intelligence and intelligence testing remains contentious.

To the extent that we see value in assessing intellectual competencies of individuals and groups, then continuing the search for greater understanding of the construct is a worthwhile endeavor. In this society, intelligence test scores have been used not only to make selection and placement decisions but also to predict how individuals will perform in the future or on other tests. The more theoretically grounded the construct, the greater our faith in the results from its assessment. Thus, for very practical reasons, the theoretical basis of intelligence needs to be determined. The chapter begins with a more elaborate discussion of a rationale for a conceptually driven assessment of intelligence. Next, criteria for a theory of intelligence that would inform a particular perspective of assessment are set forth. Philosophical, subjective, and objective conceptions of intelligence are advanced and evaluated against these criteria. The chapter ends with an identification of a number of theoretical propositions that form the conceptual basis for an assessment system of intelligence.

Rationale for a Conceptually Driven Assessment of Intelligence

Assessment is a process of gathering evidence of an individual's performance on a given task and of making interpretations and inferences about behavior on the basis of such evidence for a variety of purposes. Intelligence or IQ tests (e.g., the Wechsler scales) represent one form of assessment and are designed to enable systematic observation of individual and group differences in intellectual performance in a controlled setting. Biological or environmental influences are often used to account for differential performances, and purposes served by test data include decision making about selection, placement, and prediction about certain types of school achievement and job performance. For example, test scores are used, in part, to determine eligibility for gifted, mentally retarded, and special education programs and to predict who will do very well or poorly academically as measured by academic achievement tests. For the past three decades, proponents and critics have argued about the appropriateness of IQ tests for these purposes and the scientific legitimacy of biological or environmental interpretations for observed difference in performance. Given the fact that data from IQ tests are used, in part, to make decisions with respect to allocation of the nation's resources and opportunities, it is understandable why the IQ controversy has generated such strong emotional debate. Because of the high "stakes" consequences of intelligence and the irresolution of the nature-nurture question, we think that any measure purporting to measure such an enigmatic construct should be grounded in theory. Furthermore, we contend that any effort to assess intelligence should first inform diagnostic and prescriptive decision making. Only when decision makers have a broad and deep understanding of the sources of observed differences in intellectual performance and the prescriptive utility of such information should considerations about selection and prediction be explored.

What, then, are some reasonable criteria for any perspective of intelligence? This is a difficult question because the construct itself is a cultural invention created for the purpose of appraising who has how much of it according to the value system of any given society at any point in time! It is quite possible that a researcher or test designer may have a particular view of intelligence and can systematically observe behavior in task performance purporting to embody the construct. What counts for evidence, however, and how much is enough are questions for which there can be endless debate. We share Gardner's (1983) view that two prerequisites of a theory of intelligence should be that (a) it captures a sufficiently broad range of abilities that are genuinely useful and important at least in certain cultural contexts and (b) it should be verifiable.

Criteria for Conceptions of Intelligence

Our review of the literature about intelligence testing and our own experiences in this area during the past 10 years suggest that when these prerequisite conditions are met, there are at the very minimum the following four criteria that any conception of intelligence should meet:

1. Operational definition of the nature of the mental activity in tasks defined as "intellectual": To make an accurate determination of individual and group strengths and weaknesses, precise definition of the nature of the mental operations or cognitions underlying performance indicative of intelligence needs to be made.
2. Identification of a symbol system on which cognition(s) operates: Information is represented through language, numbers, pictures, gestures, or other types of symbol systems. The meanings and knowledge an individual acquires depend, in part, on the efficiency of operation of cognitions on any given symbol system. Explanations of differences in observed performance on intellectual tasks must take into account the symbol system on which cognition must operate.
3. Specification of the experiences within the context relevant for intellectual functioning: The kinds of experiences to which individuals are exposed may constrain or enable the efficiency and accuracy of the deployment of cognition. Because all experiences unfold within a particular setting or context, explanations of differences in observed performance on intellectual tasks must consider both the experiential and contextual realms within which cognitions are embedded.
4. Specification of the degree of susceptibility of cognition to biological and environmental influences: If prognosis for the future functioning of the individual is favorable, then the responsiveness of the cognitions to training or environmental manipulation needs to be ascertained. If, however, prognosis for future intellectual behavior is unfavorable, then the resistance of the cognitions to external stimulation would also need to be determined.

These are the criteria that we use to examine the philosophical, subjective, and objective conceptions of intelligence.

Philosophical Conceptions

Although the term *intelligence* had not been widely used until the seminal work of Binet and Simon in 1905, philosophical conceptions of the mind, intellect, the soul, and rationality can be traced back to ancient Greece. Robinson (1994) reviewed the evolution of philosophical views of intelligence from ancient through medieval and modern periods and found some consistency in the thinking of many philosophers regarding the nature of intelligence. For example, from examination of Plato's dialogues, Robinson (1994) found that Plato conceived of the intellect as an innate and god-given construct and equated it with wisdom-knowledge of universally true and unchanging principles. Moreover, it is to be distinguished from mere craft or skill that is acquired through sense-based experiences. For Plato, acquisition of wisdom used interchangeably with pure rationality could only be attained through a lifetime of the right kind of philosophical reflection and contemplation. Aristotle, Plato's star pupil, shared a similar view of intelligence as the rational awareness of universal principles to be distinguished from factual knowledge acquired through experience.

In medieval times, as Robinson (1994) pointed out, conceptions of intelligence were consistent with those of the ancient Greek philosophers. For example, Thomas Acquinas (1225-1274), in an effort to distinguish rote memory and sensory-based knowledge from cognitive abstractions, proposed a two-process perspective of intelligence: a passive intellect acted on by sensory information and an active intellect that engages the passive intellect in ways that enable the discernment of universal principles.

In the 16th century, Juan Huarte de San Juan, a Spanish physician and scholar, considered intelligence as a three-faceted construct: As reported by Linden and Linden (1968), Huarte described intelligence as (a) docility in learning from a master, (b) understanding and independence of judgment, and (c) inspiration without extravagance. It is claimed (Franzbach, 1965) that this characterization of intelligence influenced the thinking of German philosophers during the 18th and 19th centuries.

During the 17th and 18th centuries, according to Robinson (1994), three distinct conceptions of intelligence were advanced. The first one was rationalistic and was closely associated with the ideas of René Descartes and Gottfried Wilhelm von Leibnitz. Both, in the tradition of the Greek philosophers, held the view that intelligence was the capacity to comprehend abstract principles, and they argued for the prior existence of the intellect within the individual to make intelligible the multitude of experiences perceived through sensory stimulation. The second perspective—the empiricist—was associated with the writings of John Locke and David Hume who, using the principle of association, proposed that all knowledge can be reducible to objects as experienced through the senses. The third perspective was a biological one put forth by a number of philosophers who based their views on the assumption of a relationship of the organization of an animal's body and its adaptive abilities. According to this view, intelligence is defined as the problem-solving abilities made possible by the degree of efficiency of this organization.

It was not until the 19th century, however, with the publication in 1859 of Darwin's *Origin of the Species*, that investigations concerning the nature of intelligence became more pragmatic and less philosophical. Two provocative notions emerged from Darwin's work that were later explored by his half-cousin, Francis Galton. First is the proposition that the development of intelligence over the life span may, in some respects, resemble the development of intelligence from lower to higher species. Second is the assumption that development is continuous and therefore human beings could be subjected to the same type of scientific investigations as those conducted with animals.

Ten years later, Galton, in *Hereditary Genius* (1869), suggested that intellectual genius tends to run in families and is therefore inherited. In one of the chapters, "Classification of Men According to Their Natural Gifts," he alluded to the presence of specific and general abilities as the basis for individual differences in intelligence. In 1883, he published *Inquiries Into the Human Faculty and Its Development Among Human Beings* in which he explored the difficulties in measuring mental aptitudes and proposed a series of psychophysiological measures to discriminate the mentally strong from the mentally weak. From 1884 to 1889, he sought empirical validation of these claims using a variety of simple physical and sensory tasks at his anthrometric laboratory.

Commentary

None of the philosophical conceptions of intelligence met the prerequisites or criteria very well. All described intelligence as a mental activity, although the nature of the cognition differed. The ancient Greeks, the early Christians, and the rationalists considered that the capacity to frame and understand universal or abstract principles was an essential quality of the nature of mental life. The empiricists, however, proposed a theory of the mind as a collection of elementary sensation, combining to form simple to complex ideas through the principle of association. Although the issue of experience was discussed, its role in intellectual functioning differed. The Greeks, the early Christians, and the rationalists argued that although experience may enable rote memory and sense-based knowledge, these cannot account for the capacity to acquire knowledge of universal principles—the essence of intelligence. In contrast, the empiricists believed that all knowledge is reducible to stimuli acquired through the senses. None gave explicit attention to the role of a symbol system. With regard to the criterion pertaining to biological or environmental influences, the Greek scholars believed that intelligence was an innate and god-given construct. Some modern-day scholars, although not attributing divine blessing to those who possess it, believed that the genesis of intelligence was located in the organization of the body and nervous system. Although these conceptions may have met the prerequisite criterion of being useful or important in some cultural contexts, they did not meet the other prerequisite of a broad range of human abilities. Perhaps the greatest threat to these conceptions of intelligence, however, is the absence of verifiable evidence about its nature.

Subjective Conceptions of Intelligence

Subjective views of intelligence are based on people's commonsense theorizing or beliefs about the nature of intelligence. Unlike the methodologies used to study objective conceptions of intelligence, research uses self-report measures with a Likert-type format to elicit peoples' stated positions regarding intellectual functioning. Speculating that the construct was no more than society's cultural invention of it as a prototype of what its members value as a culture, Neisser (1979) used this approach to examine laypersons' conceptions of intelligence. These findings confirmed his earlier distinction between academic (used in classroom tasks) and practical (used in tasks people face in their everyday lives).

Many cross-cultural studies have found variations within and among cultural groups in their conceptions of intelligence. For example, Wober (1972) found that Ugandan teachers and groups influenced by Western ideas characterized intelligence in terms of speed, whereas the Ugandan villagers associated intelligence with words such as careful, active, and slow. In a later study of views of intelligence from different tribes in Uganda, Wober (1974) found that in Batoro tribes, people associate intelligence with terms such as obedient, yielding, and soft, whereas members of the Baganda tribe associated intelligence with words such as persistent and hard.

Super (as cited in Sternberger, 1985c) found that the concept of intelligence had different meaning when applied to children and adults among Kokwet people in western Kenya. For example, the connotation of cleverness, inventiveness, and unselfishness was associated with the word *utat* when applied to adults. The connotation of fluency and quickness in verbal comprehension on complex tasks and interpersonal skills was associated with the term *ngom* when applied to children.

Sternberg, Conway, Ketron, and Bernstein (1981) asked samples of experts in the field of intelligence and samples of laypersons in a train station, library, and supermarket to define and rate characteristics of intelligent people. Both groups had similar conceptions of intelligence that distinguished between competent performance on practical and academic tasks. Finally, Okagaki and Sternberg (1993) compared views of immigrant parents from Cambodia, Mexico, the Philippines, and native-born Anglo-Americans and Mexican Americans regarding their conceptions of intelligence of a typical first-grade child. They found that, with the exception of Anglo-Americans, all other groups identified intelligent characteristics, such as social skills, practical school skills, and motivation, as equally or more important than cognitive characteristics.

Commentary

Subjective conceptions of intelligence met the prerequisites and some criteria but not others. On the positive side, these conceptions define a very broad array of cognitions, spanning those used in school-like tasks and in a real-world environment. All these views considered experiences and context as important dimensions of intelligence. Lack of precision in terms of the actual behaviors indicative of these cognitions, however, renders judgments of them tentative at best. Consensus among the views of laypersons and experts in the area of intelligence research gives some validity to these conceptions. None of the conceptions addressed the criterion regarding the susceptibility of cognition to biological influences.

Objective Conceptions of Intelligence

Objective conceptions of intelligence are derived from findings from observation of individual differences in performance on intellectual tasks. Methodological developments have influenced these perspectives in the 20th century, including factor analytic and information-processing models and standard psychometric and information-processing tests of mental abilities. In addition, clinical studies in neuropsychology and field research in anthropology and psycholinguistics using observational methods and conversational analysis are shedding new light to an old question. In the following sections, we examine some of the more common conceptions of intelligences that have used these procedures to study individual differences in behavior indicative of intelligence.

Psychometric Theories of Intelligence

Psychometric theories of intelligence seek an understanding of intelligence in terms of the way it is measured through the use of a statistical-mathematical technique called factor analysis. It involves the examination of a matrix of intercorrelations or covariances for a set of cognitive tasks (most often standardized scores on psychometric tests of mental ability) to uncover common patterns of individual differences in performance on these tasks. The observed patterns or factors sometimes referred to as latent traits are presumed to be manifestations of individual differences in cognitive abilities or potentials as hypothesized by the theorist. The cognitive tasks used to identify individual differences in cognitive abilities are similar in many respects to the ones initially developed by Binet and Simon in 1905. Successive factorization of correlation matrices yields different numbers of factors at varying levels of generality, and such information is used to infer the structure or organization of the human intellect. Consider some examples of conceptions of intelligence with varying factors and structures that have been widely cited in psychology texts.

Spearman

Charles Spearman (1927) was among the first to propose a two-factor theory of intelligence—a general factor and specific factors. The general factor, which he termed *g*, is an index of general mental ability that accounts for the patterns of intercorrelations observed among performance on various cognitive tasks or intelligence tests. Specific factors that underlie only a single task or test are indicative of specific abilities. He believed that the general factor of intelligence reflected individual differences in three mental processes: "apprehension of experience"—perceiving and understanding the task of interest; "eduction of relations"—inferring some logical abstraction such as a similarity or comparison from two or more stimuli; and "eduction of correlates"—finding a second idea logically related to a previously stated one. (See *The Nature of "Intelligence"* and the *Principles of Cognition* for a detailed discussion of these attributes of intelligence.)

He also speculated that individual differences in these mental abilities, particularly the ability to educe relations and correlates, may be explained in terms of mental energy or power that might have a physiological base.

Thurstone

Thurstone (1938) proposed a seven-factor theory of intelligence. It is different from the one common factor in Spearman's model in that he contends that seven common factors describe all the intercorrelations observed on performance on cognitive tasks or tests. Thus, for Thurstone, seven primary abilities are represented by these factors. The nature and content of these abilities are described as follows:

Verbal comprehension: the ability to demonstrate knowledge of vocabulary and comprehension skills

Verbal fluency: the ability to generate rapid production of words quickly and in a limited time

Number facility: the ability to do computational arithmetic and reason quantitatively

Spatial visualization: the ability to mentally manipulate geometric designs making same-different comparisons

Memory: the ability to remember associations with words and pictures

Reasoning: the ability to deduce and induce logical abstractions from concepts or symbols

Perceptual speed: the ability to recognize quickly symbols embedded among an array of other symbols

Wechsler

Wechsler (1944), like Spearman, conceived of intelligence as a global or aggregate capacity to think rationally, act purposefully, and cope effectively with the environment. The Wechsler scales (for adults, children, and preschoolers) are based on this conception of intelligence. Despite the general nature of the construct, Wechsler believed that intelligence can be expressed in various ways. It is for this reason that he constructed his scales into two domains, verbal and performance, with many subtests within each domain. The verbal subtests assess verbal comprehension, verbal concept formation, and verbal reasoning, and the performance subtests assess perceptual organizational abilities, visual and motor abilities, and abstract nonverbal reasoning. Thus, the scales yield two scores—verbal and performance IQ—in addition to a full scale IQ. He did not equate these IQ scores with intelligence because he maintained that other nonintellective factors (persistence, drive, and motivation) were involved in intelligent behavior. No specific quantitative index of these nonintellective factors accompanied the IQ scores, however. As a clinician, he advocated weighing both the intellective and nonintellective factors in the assessment of intelligence.

Guilford

Guilford (1967) conceived of intelligence as comprising three components: the mental operation, the content of the material, and the

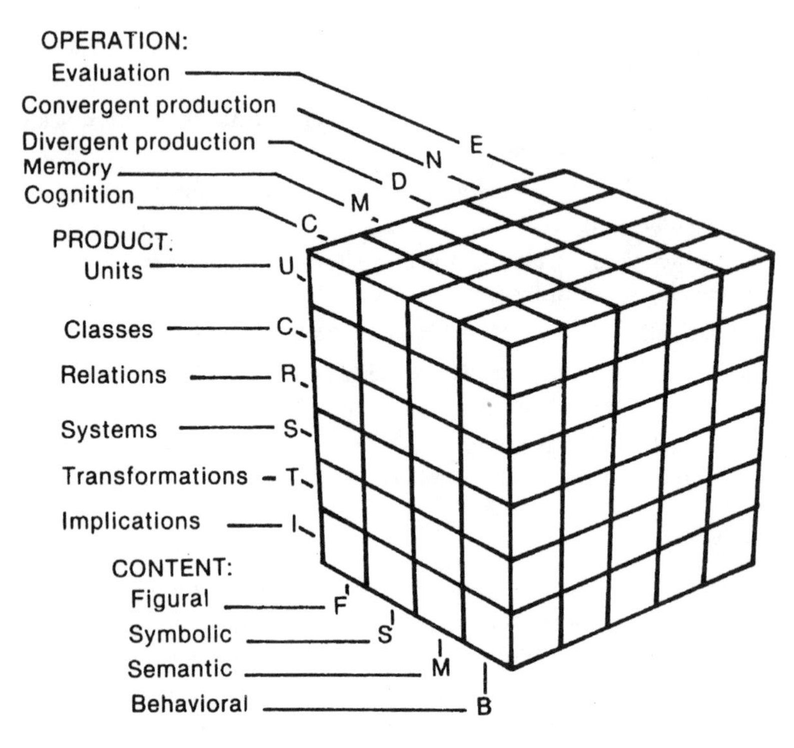

Figure 2.1. Guilford's structure of intellect model from B. B. Wolman (1985), *Handbook of Intelligence: Theories, Measurements, and Applications.* Reprinted by permission of John Wiley & Sons.

product. There are five kinds of operations: evaluation, convergent production, divergent production, memory, and cognition. There are four types of content: figural, symbolic, semantic, and behavioral. There are six kinds of products: units, classes, relations, systems, transformations, and implications. These components are independently defined so that, in combination, the four contents, five operations, and six products can produce 120 three-way combinations of mental abilities. The cubic structure of the intellect is shown in Figure 2.1.

Abilities manifested by the performance on certain cognitive tasks depended on the manner in which they were presumed to sample facets of operations, content, and product. Guilford (1967) claimed that it was possible to arrange these primary factors into a distinctive model popularly known as the Structure of the Intellect model. Similar to Thurnstone but unlike Spearman, he did not share a unitary conception of intelligence.

Horn

Horn (1991a) proposed that human intelligence comprises at least nine broad Gf-Gc (general/fluid-general/crystallized) abilities. For the past 25 years, studies indicating covariability among abilities have led to the conception of the theory as a system of factors among factors. Initially, however, the theory was based on the seminal work of Cattell (1941, 1943), who argued that general intellectual ability was of two types: fluid ability and crystallized ability. Fluid ability was conceived as being able to flow into many diverse types of mental activities. In contrast, crystallized ability was presumed to underlie the end product of an individual's exposure to education, training, or other types of experiences. Fluid ability was presumed to consist of basic reasoning and related higher-order mental processes particularly evident in novel situations. Crystallized ability, however, was presumed to reflect the ability of an individual, partly on the basis of fluid ability, to learn and benefit from exposure to experiences within one's culture. Over the years, the theory has undergone change as new evidence from numerous studies (Cattell & Horn, 1978; Gustafsson, 1984; Horn, 1965; Woodcock, 1990) revealed seven other factors in addition to the ones indicative of fluid and crystallized ability. Nonetheless, the term Gf-Gc has been retained and is associated with the work of both Horn and Cattell. As depicted in Figure 2.2, the structure of the intellect is hierarchical. Consider the nature and content of the following nine broad dimensions of the Horn-Cattell Gf-Gc theory:

Fluid intelligence: the ability to understand relations among stimuli and to make inferences and to understand implications between and among stimuli particularly with complex novel tasks

Crystallized intelligence: the ability to acquire the breadth and depth of knowledge from the dominant culture

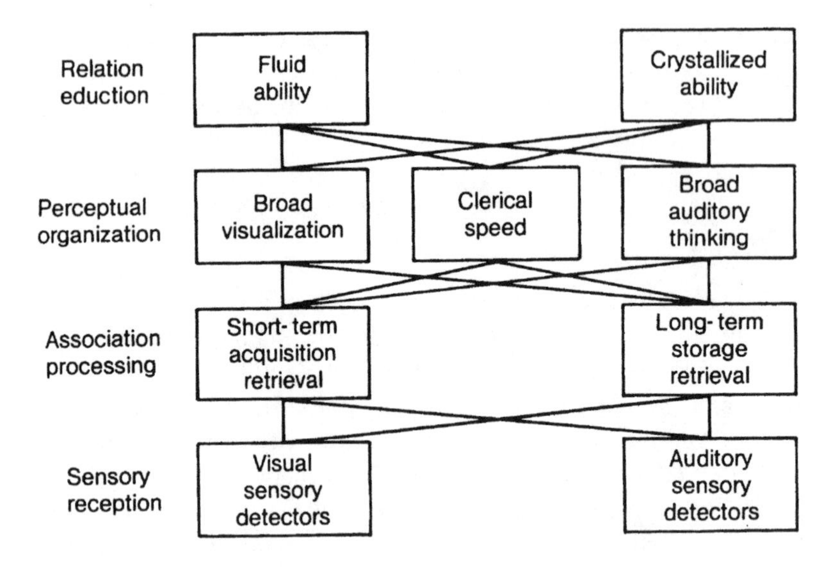

Figure 2.2. Horn's Hierarchical Model of Intelligence (Simplified Version) from B. B. Wolman (1985), *Handbook of Intelligence: Theories, Measurements, and Applications.* Reprinted by permission of John Wiley & Sons.

Quantitative ability: the ability to use quantitative information and to manipulate number symbols

Long-term storage and retrieval: the ability to store information in long-term memory over a long period of time (minutes, hours, weeks, and years) and to fluently retrieve it later through association

Short-term memory: the ability to maintain awareness of information and to recall it within a few seconds

Processing speed: the ability to quickly scan and respond to simple but timed tasks

Correct decision speed: the ability to quickly decide and respond accurately to tasks of moderate difficulty

Auditory processing: the ability to perceive sound patterns, to maintain awareness of order and rhythm among sounds under distortion or distraction, and to understand relationships between and among different groups of sounds

Visual processing: the ability to perceive and manipulate symbols of varying shape and to identify varying spatial configuration of them

Carroll

In a recent review and synthesis of factor analytic studies, Carroll (1993) proposed a three-stratum theory that involves the classification of abilities according to the generality of factors across domains of cognitive performances or tasks as well as the level and speed by which these tasks or performances are to be differentiated. The abilities located at each of the strata may be categorized as narrow (Stratum 1), broad (Stratum 2), and general (Stratum 3). Narrow, first-stratum abilities are indicative of a large number of specialized abilities presumed to reflect the effects of experience and learning. Broad, second-stratum abilities are indicative of moderate specialization of abilities and are particularly useful for understanding the breadth and scope of human cognition (fluid intelligence, crystallized intelligence, general memory and learning, broad visual perception, broad auditory perception, broad retrieval ability, broad cognitive speediness, and processing speed). According to Carroll (1993), these abilities represent "basic constitutional and long-standing characteristics of individuals that can govern or influence a great variety of behaviors in a given domain" (p. 634). As shown in Figure 2.3, general, third-stratum ability is indicative of general intelligence that reflects the domination of the second-order factors by a third factor.

Intelligence as a Multidimensional Construct

Conceptions of intelligence as a multidimensional construct have emerged in recent years. Theorists who subscribe to this view consider the internal and external world of the individual. Although cognitive processes represent an important characteristic of intelligence, the kinds of experiences in which these cognitive processes are developed and displayed and in what kinds of context are also important components of intelligence. Three theorists who share this view of intelligence are presented.

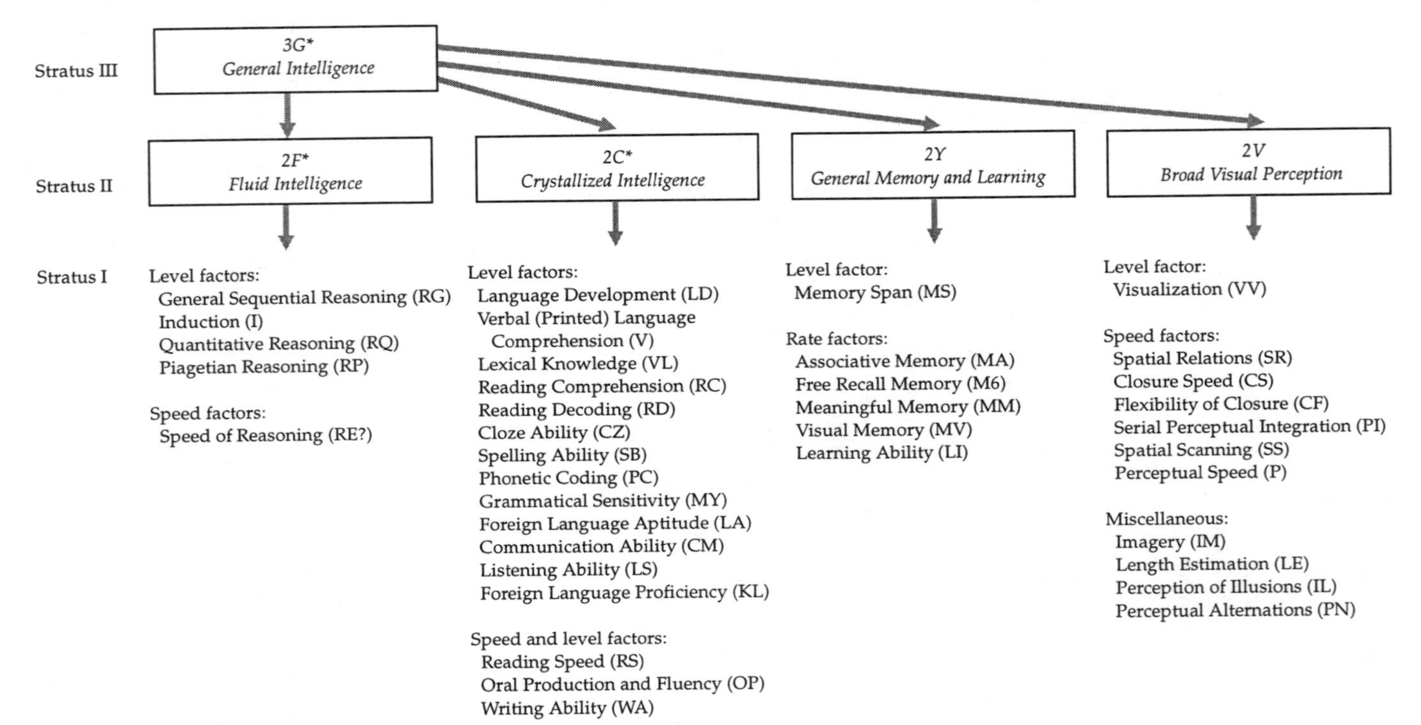
Stratus III
3G*
General Intelligence
Stratus II
2F*
Fluid Intelligence
2C*
Crystallized Intelligence
2Y
General Memory and Learning
2V
Broad Visual Perception
Stratus I
Level factors:
General Sequential Reasoning (RG)
Induction (I)
Quantitative Reasoning (RQ)
Piagetian Reasoning (RP)
Speed factors:
Speed of Reasoning (RE?)
Level factors:
Language Development (LD)
Verbal (Printed) Language Comprehension (V)
Lexical Knowledge (VL)
Reading Comprehension (RC)
Reading Decoding (RD)
Cloze Ability (CZ)
Spelling Ability (SB)
Phonetic Coding (PC)
Grammatical Sensitivity (MY)
Foreign Language Aptitude (LA)
Communication Ability (CM)
Listening Ability (LS)
Foreign Language Proficiency (KL)
Speed and level factors:
Reading Speed (RS)
Oral Production and Fluency (OP)
Writing Ability (WA)
Level factor:
Memory Span (MS)
Rate factors:
Associative Memory (MA)
Free Recall Memory (M6)
Meaningful Memory (MM)
Visual Memory (MV)
Learning Ability (LI)
Level factor:
Visualization (VV)
Speed factors:
Spatial Relations (SR)
Closure Speed (CS)
Flexibility of Closure (CF)
Serial Perceptual Integration (PI)
Spatial Scanning (SS)
Perceptual Speed (P)
Miscellaneous:
Imagery (IM)
Length Estimation (LE)
Perception of Illusions (IL)
Perceptual Alternations (PN)

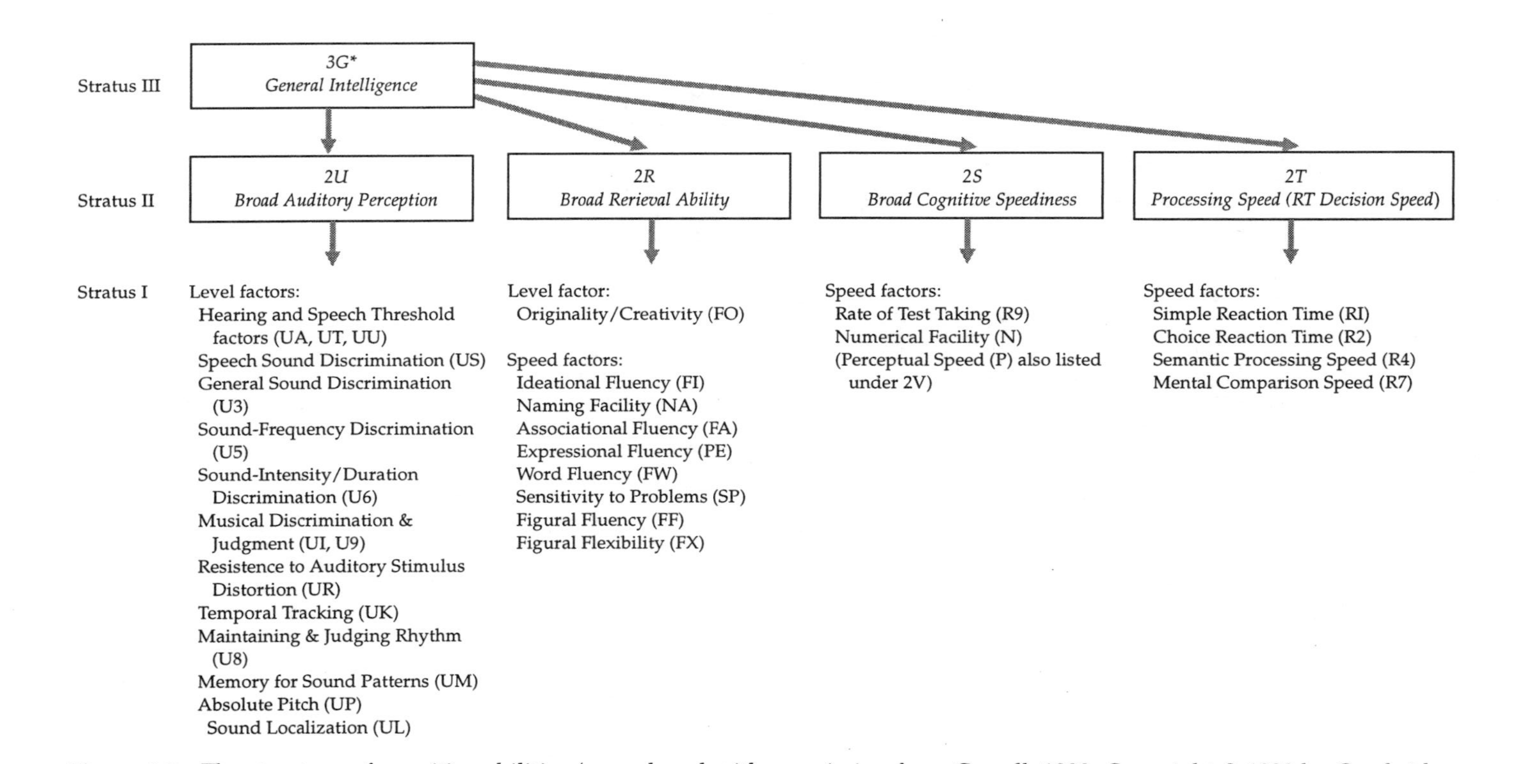

Figure 2.3. The structure of cognitive abilities (reproduced with permission from Carroll, 1993. Copyright © 1993 by Cambridge University Press.

***In many analyses, factors 2F and 2C cannot be distinguished: they are represented, however, by a factor designated 2H, a combination of 2F and 2C.**

Sternberg's Triarchic Theory

Robert Sternberg (1985c) conceives of intelligence as a triarchic construct involving three interrelated subtheories; first, a componential theory that focuses on the cognitive processes that the individual uses in the performance of intellectual tasks. Sternberg distinguished three classes of processes: (a) metacomponents—general, executive-like processes used to plan, monitor, and evaluate one's performance on a task; (b) performance components—specific processes used in the actual performance of the task; and (c) knowledge-acquisition processes—specific processes used in the learning of new words. Together these processes form the bases for the other contextual and experiential subtheories. Second, according to Sternberg, the contextual subtheory describes the purposeful use of cognitive processes in adaptation to, selection, and shaping of a real-world context important to one's life. Third, the experiential subtheory posits that the individual may use these cognitive processes in situations that require the ability to deal with novel kinds of tasks and the ability to automatize the processing of information. For Sternberg, the basic unit of analysis for explaining individual differences in intellectual functioning is the cognitive process or component. A task or situation, however, is said to measure intelligence to the extent to which processes are deployed in contextually meaningful situations that are relatively novel or to which individuals respond with automaticity. Components of intelligence and the triarchic theory are illustrated in Figures 2.4 and 2.5, respectively.

Ceci's Bioecological Treatise on Intellectual Development

Stephen Ceci's perspective on intelligence is both developmental and contextual. For Ceci (1990), a child is born with a number of biologically constrained cognitive muscles (the biological architecture to remember, to develop expectancies, classify, etc.) that are moderated by certain factors within the ecology in which the child grows and functions. These experiences within the environment to which the child is exposed set in motion biologically constrained potentials along multiple pathways, each of which has relevance for

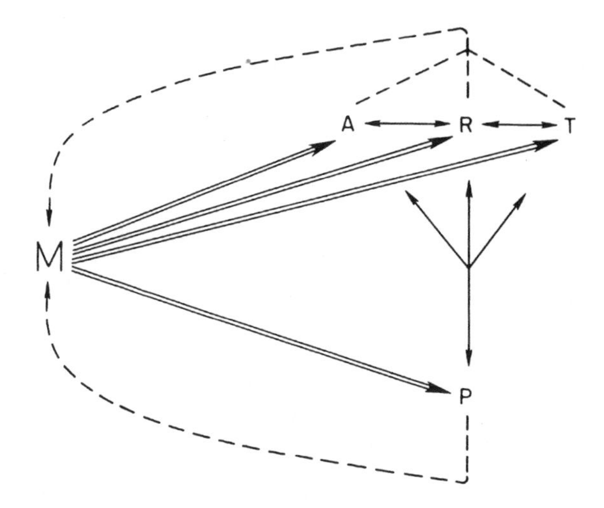

Figure 2.4. Components of intelligence. Interrelations among components serving different functions. *M*, a set of metacomponents; *A*, *R*, and *T*, a set of knowledge acquisition components as they function in the acquisition (*A*), retrieval (*R*), and transfer (*T*) of information; *P*, a set of performance components. Direct activation of one kind of component by another is represented by solid double arrows. Indirect activation of one kind of component by another is represented by single solid arrows. Direct feedback from one kind of component to another is represented by single broken arrows. Indirect feedback from one kind of component to another proceeds from and to the same components as does indirect activation and so is shown by the single solid arrows (reproduced with permission from Sternberg, 1980, p. 578. Copyright © 1980 by Cambridge University Press).

the crystallization of intellectual competencies that form the basis of adult intelligence. Thus, for Ceci, there are four facets to the development of intelligence: (a) multiple intellectual potentials; (b) context—an ecology that can either support or constrain the development and expression of the intellectual potentials; (c) domain-specific knowledge—the amount and quality of knowledge that can positively or negatively influence the efficiency and accuracy of the use of these intellectual processes; and (d) appropriate elicitors must be present within the ecology to motivate the development and expression of these intellectual potentials.

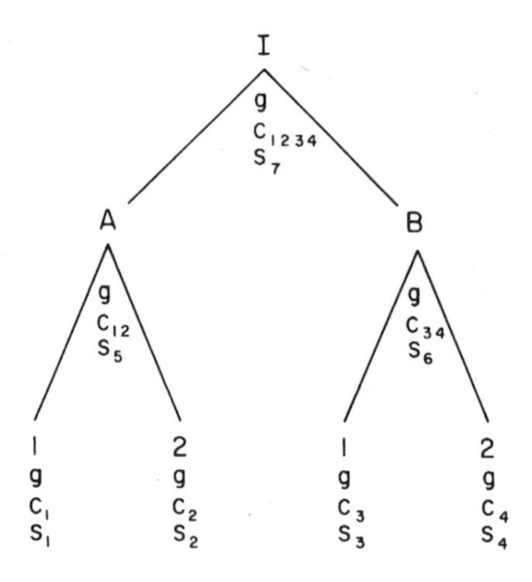

Figure 2.5. The triarchic theory: subtheories. Interrelations among components of different levels of generality. Each node of the hierarchy contains a task, which is designated by a roman or arabic numeral or by a letter. Each task comprises a set of components at the general (*g*), class (*c*), and specific (*s*) levels. Abbreviations used: *g*, a set of general components; c_i and c_j each refer to a set of class components; c_{ij} refers to a concatenaged set of class components that includes the class components from both c_i and c_j; s_i refers to a set of specific components (reproduced with permission from Sternberg, 1980, p. 579. Copyright © 1980 by Cambridge University Press).

Gardner's Theory of Multiple Intelligences

Howard Gardner is among a growing number of psychologists to argue that the human mind is quite modular in design and that separate and independent cognitive processes seem to underlie the performance on intellectual tasks. On the basis of his own studies of the development and breakdown of cognitive- and symbols-using capacities as well as those from the neuropsychological literature, Gardner and colleagues (Gardner, Howard, & Perkins, 1974; Gardner & Wolf, 1983) hold the view that an individual's precocity with one form of symbol use may not necessarily carry over to other forms. That is, individuals use different cognitive processes when engaged in tasks involving numerical, pictorial, linguistic, gestural, and other kinds of symbolic systems. He elaborated on this cognition and symbol system relationship in his 1983 *Frames of Mind*. Here, he coined the concept of multiple intelligences and proposed that individuals are capable of intellectual functioning in at least seven relatively autonomous areas and that they may have strengths or weaknesses in one or several areas. These multiple intelligences are logical-mathematical, linguistic, musical, spatial, bodily kinesthetic, interpersonal, and intrapersonal. Gardner believed that environmentally enriched contexts are likely to enable the nurturance of multiple emerging capacities.

Perspectives of Intelligence as Situated Cognition

The conception of intelligence as situated in its historical, social, cultural, and physical environment is shared by researchers from diverse disciplines, including anthropology, sociology, psychology, and psycholinguistics. A core feature of the situationist view is the reciprocal nature of the relationship between the person and the environment such that one cannot study cognition disembedded from the context in which it develops and finds expression. Using methodologies from traditions such as critical theory, ethnography, and conversational analysis, researchers examine human behavior in natural settings: the individual's interactions with others, the specifics of the person's situation, as well as the larger cultural and historical forces that embody these microlevel activities.

Vygotsky

Lev Vygotsky studied human cognitive functioning from a developmental perspective that holds that sociocultural processes are key influences on development. He makes a conceptual distinction between two classes of cognitive processes: lower (natural) and higher (cultural). He considered elementary perception, memory, and attention as rudimentary processes that create a biological predisposition for the child's development. According to Vygotsky, these lower cognitive processes become transformed into higher cognitive processes, such as reasoning, planning, and logical memory, through mediated activity and psychological tools. Three assumptions guided his work: (a) The individual cognitive processes have their origin in social interaction; (b) the zone of proximal development (the difference in performance with and without adult guidance or in collaboration with more capable peers) is one of the primary mechanisms that accounts for the development of individual cognitive functioning; and (c) the development of individual cognitive processes is mediated by tools, language, signs, or other symbolic systems. Thus, for Vygotsky, intellectual ability is not a natural entity but a sociocultural phenomenon that emerges through social mediation.

Boykin

A. Wade Boykin is among a growing number of researchers in racial and ethnic minority psychology to make the case for the relationship between cultural contexts and cognitive performance. For Boykin and colleagues, cognitive performance is inextricably wedded to the individual's views and values about what tasks should and would want to be performed. His research program is conceptually anchored within a framework of cultural integrity as it pertains to African American culture and children's cognitive performance. For Boykin, African American culture is characterized by three nonoverlapping realms of experience: mainstream, minority, and Afrocultural. Mainstream experience describes the beliefs, values, and behavioral styles common to most members of the U.S. society. Minority experience denotes certain defense mechanisms or coping strategies developed by members of ethnic and linguistic minority groups within a dominant society. Afrocultural experience refers to the beliefs, values, and behavioral styles of African descendants throughout the diaspora and traditional West African worldviews. The conflicts created by these three divergent psychological realities at once create what Boykin (1986) calls a "triple quandary" for most African Americans: "They are incompletely socialized to the Euro-American cultural system; they are victimized by racial and economic oppression; they participate in a culture that is sharply at odds with mainstream ideology" (p. 66).

For more than 15 years, Boykin and colleagues (Allen & Boykin, 1991; Boykin, 1977, 1979, 1982a; Boykin & Allen, 1988; Boykin, DeBritto, & Davis, 1984; Boykin & Toms, 1985) have engaged in research that examined the degree of success and failure of African American children on cognitive tasks in contexts reflective of these multiple experiences.

Lave

Since the early 1970s, the research investigations of Lave and colleagues have provided consistent evidence for the conception of cognition as a situated sociocultural process. Examples of this position include her observations of cognitive asymmetry in the performance of Liberian tailors on school-based and non-school-based mathematics (1977) and the context specificity of cognition in everyday mathematics tasks among grocery shoppers (1988). Moreover, observations of the opportunities afforded by certain cultures to involve children in the mature practices of their communities have added ecological validity to the notion of intelligence as a situated cognitive process. Collectively, from these works she has developed a theoretical account for the interdependence of learning and cognition as individuals become members of their communities of practice.

Commentary

For ease of discussion of the objective conceptions of intelligence in relation to the criteria, we outline each of the criteria and consider the extent to which they were met by the various perspectives. The goal here is not to do an exhaustive analysis of each perspective but rather to determine which among them offers the best guidance for diagnostic and prescriptive assessment.

Operational Definition of Cognition(s) Underlying Performance of Intellectual Tasks

Psychometric conceptions of intelligence have a vast amount of evidence in support of the mental operations involved in tasks purporting to measure intelligence. These perspectives all seem to share the view that the mental activity underlying the performance on certain types of tasks involves the capacity to perceive, to remember, to reason, and to acquire knowledge efficiently. What seems to be the point of contention, however, is to determine the relative importance of these underlying cognitions in defining the essence of intelligence and to agree on a structure of the mind that best represents this essence. Conceptually, to the extent that these cognitions have differential meaning in different cultural contexts and the extent to which differential standards are expected of individuals or groups who execute them, there will continue to be differences in its conceptions. A similar claim can be made methodologically as well. Differences are also likely to occur to the extent that there is no consensus regarding a prior criteria for use of factor analytic techniques in the analysis and interpretation of observed differences on intellectual tasks. These caveats make more understandable the seemingly endless variations in the types of cognitions underlying intelligent behavior (e.g., Carroll's, Spearman's, and Wechsler's hierarchical structure of general intelligence; Thurstone's unordered arrangements; Horn's and Cattell's second-order system; and Guilford's cubic organization of multiple intelligences).

Despite the agreement among psychometric theorists regarding the mental operations underlying the performance on intellectual tasks, there is no agreed on understanding of the number and organization of these cognitions. Individual tasks of perception, memory, reasoning, comprehension, and knowledge are more likely to be less complex than tasks requiring different combinations of cognition. Even within particular cognitions there is variation in terms of how they cluster in a given universe of tasks. For example, Guilford's structure of the intellect and Thurnstone's primary abilities represent different combinations of the same cognitions. Also, Carroll's three-stratum model and Horn-Cattell Gf-Gc models identify similar cognitions, but Carroll's conceives of intelligence as involving narrow, broad, and general cognitions, whereas Horn and Cattell consider intelligence to consist of nine broad intellectual capabilities. These diverse ways by which different cognitions are clustered suggest that the search for the exact or precise nature of intelligence may be a fruitless one.

There is also empirical evidence and consensus for the nature of the cognitions underlying performance on intellectual tasks for both the multidimensional and the situated cognition views of intelligence. Like the psychometric perspectives, however, there is still no consensus about their exact nature.

Susceptibility of Cognitions to Biological or Environmental Influences

To our knowledge, there is no conception of intelligence that does not give some consideration to both the biological and the environmental influences on human cognition(s). What distinguishes one theorist from another is the relative weight given to biology or environment. The role of these two influences is important because either one has implications for change in subsequent intellectual functioning. To the extent that the operation of a cognition(s) is constrained by biological forces, its stable or fixed quality would render it relatively impervious to significant modification through intervention. If, however, the cognition(s) is responsive to environmental manipulation, then its labile or dynamic nature would permit modification through intervention. Many studies of the genesis of intelligence in the psychometric literature suggest that the heritability of intelligence ranges from .50 (Scarr & Carter-Salzman, 1982) to .58 (Plomin, 1985) and .70 (Bouchard, Lykken, McGue, Segal, & Tellegen, 1990). Other studies have provided moderate to strong correlational evidence of the relationship between speed or efficiency of neurophysiological processes and intellectual behavior as measured by IQ tests (e.g., Jensen, 1987, 1991; Vernon, 1990).

Other theorists (e.g., Gardner, 1983; Sternberg, 1988), however, while acknowledging the relevance of biology in intellectual functioning, seem more inclined to seek empirical evidence of the malleability of the construct and its responsiveness to training or schooling. Gardner and colleagues' work in Arts PROPEL (Gardner et al., 1989) and Project Spectrum (Krechevsky & Gardner, 1990) as well as Sternberg's (1988) program to train intellectual skills in Venezuelan schools are examples of the positive outcomes of intervention to modify cognition.

Vygotsky does not dismiss the role of heredity in behavior but seem more interested in discussing the modification of cognition through instruction. Followers in the Vygotskian tradition, other researchers (e.g., Feuerstein, Rand, & Hoffman, 1979; Lidz, 1991) have provided encouraging evidence of change in cognitive functioning through mediated interactions.

Identification of a Symbol System on Which Cognitions Operate

All conceptions of intelligence describe intellectual behavior within some symbolic domain. For example, psychometric multidimensional and situated cognition theorists examine the operation of cognition in different types of tasks requiring linguistic, spatial, numerical representation. Gardner (1983), however, considers other symbolic domains (e.g., music and dance) wherein intelligence might find expression. Also, for situation theorist Vygotsky, symbolic tools, such as signs and linguistic and mathematical systems, play a crucial mediating role in the development and expression of cognition. Although most theorists would acknowledge variation in performance as a function of the modality in which intellectual content of tasks are embedded, it is not clear how much researchers attribute differences in intellectual performance to the way individuals represent information. Ceci (1990) and Gardner (1983) consider intellectual functioning inseparable from the symbol system in which it is embedded and consequently argue for conceptually different manifestations of cognitions. Psychometric theorists (e.g., Carroll, 1993; Spearman, 1927), however, argue that the intercorrelations of intellectual tasks involving different symbol systems demonstrate the existence of a general ability.

Specification of the Context and Experiences Relevant for Intellectual Functioning

All conceptions of intelligence recognize that experiences within certain types of settings influence intellectual functioning. For example, a common and consistent finding of studies in the psychometric tradition is that there is a positive correlation between IQ scores and academic achievement (.50) and certain types of job performance (.30). One explanation for these findings is that certain types of experiences in school and in the workplace require intellectual competence as defined by IQ tests. Other researchers who argue that there is more to intelligence than what is measured on an IQ test provide evidence of experiences in nonschool settings that also require intellectual competence. For example, Sternberg, Wagner, Williams, and Horvath (1995) found that tasks that are embedded in and require everyday experiences are important for demonstrating what they call practical intelligence. Similarly, other researchers (e.g., Boykin, 1982; Ceci & Liker, 1986b; Gardner, 1983; Lave, 1988) found evidence of intellectual functioning on tasks outside the school context.

Implications of Conceptions of Intelligence for Intellectual Assessment

Some of the conceptions of intelligence reviewed in this chapter met the criteria, and others did not. Some had empirical and consensual support, and others did not. At some level, it is not surprising that all conceptions did not meet these requirements. It may be recalled that these criteria emerged as a "wish list" of attributes that we believe should guide the development or an assessment procedure that serves both diagnostic and prescriptive purposes. Second, it is an unreasonable expectation that any one theory would meet all the requirements for an assessment procedure, particularly one with such high-stakes consequences in our society. These caveats notwithstanding, the review did provide some conceptual justification for attributes associated with intelligence that do have implications for intellectual assessment. What, then, are the theoretical perspectives that met our criteria for assessment? Table 2.1 presents a list of those theoretical propositions that, when taken together, provide a more comprehensive conception of human intelligence than any single perspective.

We applied these propositions in the development of a four-tier system to assessing intelligence (see Chapter 6 for an elaboration).

Conclusion

Today, we are no more definitive in our understanding of the concept of intelligence as when the ancient Greek philosophers sought to make distinctions between its principled and sensory-based characteristics. During the past nine decades, however, the objective and subjective conceptions have broadened our knowledge of the complexity and multifacetedness of the phenomenon. In addition, methodological advances in factor analytic and information-processing

TABLE 2.1 Major Theoretical Propositions of Each Conception of Intelligence

Theorist	*Propositions on Intelligence*
Psychometric theorists (e.g., Carroll, Horn, Wechsler, and Guilford)	Cognitive abilities underlie performance on intellectual tasks
Situated cognition theorists (e.g., Boykin, Lave, and Vygotsky)	Cognitive processes have their origin in social interaction
	Emergent cognitive processes could be identified through cooperative interaction between child and adult
	Cognitive processes are developed and manifested in culturally familiar tasks and contexts
Triarchic theorist (Sternberg)	The basic unit of intelligence is the cognitive process or processes
	Novel and familiar experiences influence how cognitive processes are deployed
	Different contexts elicit different manifestations of cognitive processes
	Cognitive processes can be modified through intervention
Bioecological theorist (Ceci)	Intelligence consists of multiple cognitive processes
	The structure of knowledge influences the development and manifestation of cognitive processes
	The ecology impedes or fosters the development of biologically constrained cognitive potentials
	The ecology contains elicitors for the manifestation and development of cognitive potentials
Multiple Intelligence theorist (Gardner)	Intelligence as emerging cognitive capacities develop in multiple contexts and through multiple symbol systems

procedures and clinical and observational techniques have been useful in providing some insights with respect to the observed individual differences in performance on intellectual tasks. In the course of the review, it became apparent that experiences and the context in which these experiences are embedded play a role in intelligent behavior as do cognitive capacities that are more intrinsic to the individual. These notions about intelligence have very clear implications for diagnostic and prescriptive assessment: (a) to design measures that would identify individual strengths and weaknesses in cognitive processing as well as the contextual and experiential factors that enabled or constrained its expression and (b) to ensure that these assessment probes yield information at a sufficient level of detail to inform pedagogical planning and intervention. These issues are explored more fully in Chapter 6.

3

Culture and Cognition

Human cognition describes the mental activities that manipulate, translate, and transform information represented in any modality. Thus, for example, it can change verbal information into spatial representation or pictorial information into numerical representation. Variation in number, form, level, and organization of human cognition will depend on the density or complexity of information to be operated on. As indicated in Chapter 2, the more commonly cited cognitions involved in intellectual activities are those related to short- and long-term memory, reasoning, vocabulary, comprehension, visual processing, auditory processing, and speed of processing. We also learned that some contemporary theorists argue that these cognitions do not function in isolation but depend, in part, on certain kinds of experiences in contexts for their development and expression. In this chapter, we explore these issues further through an examination of the relationship of culture and cognition. First, various conceptions of culture are identified, and then an overview of the relevant research in this area is presented. From this review, we distill those psychologically relevant attributes of culture that systematically influence how and when cognitions are deployed in a given task or situation.

Conceptions of Culture

According to the historian Stocking (1968), the modern concept of culture emerged at the end of the 19th and the beginning of the 20th century. The research of early anthropologists through their field studies of various cultural groups throughout the world sought to understand the factors responsible for variation in thinking among various groups. This early position was in direct contrast to the views of Tylor (1874) and social scientists of the midcentury who, guided by Darwinian notions, had constructed an evolutionary model of culture. By tracing the development of culture, some societies were presumed to evolve through stages from less to more developed culminating naturally in modern culture with its technological, intellectual, and artistic traditions. This ethnocentric view was refuted, however, by other cultural anthropologists who, through systematic and prolonged observation and field notes, sought to understand variation in thinking among various cultural groups. An early popular hypothesis from this early work as represented in the work of Rivers (1926) was that different patterns of thinking were related to different environmental demands across cultures.

Geertz (1973) put forth the following definition of culture that is widely cited in cross-cultural psychology: Culture is an "historically transmitted pattern of meanings embodied in symbolic form by means of which men communicate, perpetuate, and develop their knowledge about and attitudes toward life" (p. 89). Although there was no explicit reference to thinking in this definition, Geertz's notion of the inseparability of culture and cognition of more than 20 years is widely shared by many cultural psychologists today: "The human brain is thoroughly dependent upon cultural resources for its very operation; and those resources are, consequently, not adjuncts to but constituents of mental activity" (p. 730). Gordon (1991) has extended Geertz's notion of culture to include "structured relationships, which are reflected in institutions, social status, and ways of doing things, and objects that are manufactured or created such as tools, clothing, architecture, and interpretative and representational art" (p. 101).

In an effort to portray culture as an inclusive and overarching construct in the lives of any social group, Gordon (1991) conceives it as a multidimensional construct consisting of at least five dimensions: (a) the judgmental or normative, (b) the cognitive, (c) the affective, (d) the skill, and (e) the technological. Elaborations of these dimensions may be found in Gordon and Armour-Thomas (1991).

Transdisciplinary Research on Context and Cognition

During the past three decades, numerous studies have been conducted both within and across diverse cultural groups to examine differences in performance on cognitive tasks. These studies have been guided by different perspectives about context, learning, and cognition from disciplines including anthropology, sociolinguistics, education, psychology, and sociology. A common research inquiry across disciplines was the search for an explanation for the finding of context-specific cognition within and between individuals in diverse settings and cultures. In the following sections, we present, in capsule form, a select review of the theoretical and empirical research in this area.

Anthropological and Cultural and Psychological Research

Interpreting the variable findings in the early cross-cultural Piagetian studies, Cole, Gay, Glick, and Sharp (1971) made the following observation: "Cultural differences in cognition reside more in the situations to which particular cognitive processes are applied than in the existence of a process in one cultural group, and its absence in another" (p. 233).

Today, a vast knowledge base has accumulated that is essentially consistent with Cole et al.'s (1971) insightful observation. Many of these investigations were cross-cultural in nature in Western and non-Western cultural groups on cognitive tasks of memory (Kearins, 1981; Lancy & Strathern, 1981), reasoning (Gladwin, 1971), mental arithmetic (Lave, 1977; Murtaugh, 1985), and literacy (Cole & Scribner, 1977; Rogoff & Waddell, 1982). The English publication of *Mind in Society* (Vygotsky, 1978), which laid out the propositions of the sociohistorical theory of Vygotsky and colleagues, provided a theoretical frame for understanding the observed variation in performance on cognitive tasks. A basic tent of Vygotsky's sociocultural theory is that nascent cognitive potential emerges, develops, and is displayed in a sociocultural milieu. Since then, numerous empirical studies in this tradition have sought an understanding of people's "everyday cognition" by studying their thinking in real-world tasks in multiple real-world environments.

In synthesizing this body of work, Rogoff and Chavajay (1995) identified the following key assumptions common to disciplines that use a sociocultural approach to study differences in cognitive performance:

1. The use of the concept of activity as the unit of analysis to examine human cognition in tasks of a sociocultural nature
2. The dual analysis of development and cognitive process
3. Analysis of performance that integrates cognitive processes at the individual, interpersonal, and community level
4. The study of differences and similarities in performance
5. The research methods as tools in the service of research
6. The historical and cultural embeddedness of the research question itself

Cognitive, Experimental, and Psychological Research

Within the past 15 years, empirical research findings in cognitive and experimental psychology have been consistent with those of anthropology and cultural psychology. Using a different theoretical lens, studies were based on the assumption that intelligence is more than what an IQ test measures. More specifically, researchers searched for proof of the elusive theory that is presumed to explain individual differences in performance on standardized measures of intelligence. For example, Sternberg (1988) has argued that an important aspect of intelligence is whether mental activity is directed toward "purposive adaptation to, and selection of, real-world environments relevant to one's life" (p. 45). Labeling such intelligence in context as "practical intelligence," Sternberg and colleagues (Sternberg & Wagner, 1986; Sternberg, Wagner, & Okagaki, 1993; Sternberg, Wagner, Williams, & Horvath, 1995) have conducted a series of studies across settings and cultures. These findings provide compelling evidence that performances on measures of practical intelligence, although related to measures of performance on real-world tasks, are relatively unrelated to standardized tests of intelligence. Sternberg and colleagues attribute success on practical intelligence tasks to procedural knowledge that is tacit but that is acquired with little direct help from others but is important to the attainment of goals that people value.

Similar to the practical intelligence research as well as that from anthropology and cultural psychology, studies from experimental psychology have also found evidence of lack of cross-task correlations, even though the mental activity was isomorphic across situations or contexts (Neisser, 1979). Perhaps the most compelling evidence for the context specificity of cognitive skills is provided by the prolific work of Ceci and colleagues (Ceci, Baker, & Bronfenbrenner, 1987; Ceci & Bronfenbrenner, 1985; Ceci & Cornelius, 1989; Ceci & Liker, 1986b, 1988; Chi & Ceci, 1987).

Ceci (1990) proposed a bioecological treatise on intellectual development in accounting for the consistency of these findings. According to Ceci, the individual is born with a number of biologically constrained cognitive potentials (e.g., the capacity to classify, remember, and habituate) that are moderated by aspects of the environment, such as domain-specific knowledge and appropriate elicitors from the context for the manifestation of cognitive potentials. From this perspective, conditions within a particular context can either impede or enable the development and eventual manifestation of certain cognitive potentials. Collectively, these psychologically laden environmental variables led to his conception of intelligence as an ecologically based construct.

Research on Context-Specific Attributes and Cognition

A notion common in the previously cited research is that there are psychologically meaningful attributes within a particular context that influence the acquisition of specific types of knowledge and cognitive skills. It would appear that attributes of the task itself as well as conditions of its engagement enable or impede the individual's demonstration of knowledge and cognitive skills. Finally, knowledge and skill seem inseparable from the values and beliefs that are often implicitly reflected in the social interactions and tasks that a given culture deem appropriate or meaningful for its people. Given that we use these empirical findings as a basis for the biocultural thesis that we develop in Chapter 4, we comment again on specific studies that have examined the psychological significance of these context-specific variables on children's intellectual development.

Context. Numerous studies have examined the socialization of knowledge acquisition and cognition in various contexts, such as the home, the school, and the community. For example, some investigations focus on the various forms of language usage between significant others and children in the home in culturally and cognitively meaningful activities (Heath, 1983; Ochs & Schiefflin, 1984). Other studies examine child-rearing values and beliefs and their influence on the mother-child interactions on cognitive tasks (Guttierrez & Sameroff, 1990; Hale-Benson, 1986; Laosa, 1980; Serpell, Baker, Sonnenschein, & Hill, 1993; Steinberg, Dornbusch, & Brown, 1992). The longitudinal research of Bradley and Caldwell (1984) provides a comprehensive description of the kinds of stimuli and social interactions and structure operating within the home that are likely to influence cognitive growth and development: maternal involvement and responsivity with the child, maternal acceptance and encouragement of social competence, organization of the environment and provision of appropriate tools and material for play, and variety of stimulation in the home.

Several researchers have examined a number of factors related to schooling and intellectual development. Some studies focused on the amount of schooling, whereas others examined the quality of schooling. Specific studies are too numerous to mention here. Critical analyses of those findings, however, have been done by Ceci (1990), Nerlove and Snipper (1981), and Rogoff (1981). For example, in a recent synthesis of research on the cultural basis of cognitive development, Rogoff and Chavajay (1995) discussed the influence of schooling in fostering the following types of cognitive skills:

1. The deliberate remembering of disconnected bits of information
2. The organization of information to be remembered using taxonomic rules
3. The shifting to alternative dimensions of classification and verbalization of strategies used for their organization
4. The analysis of two-dimensional patterns through the use of media representing depth in two-dimensional drawings

Studies that have examined the role of community influence on children's intellectual development consider the parameters of children's social and intellectual work space and the opportunities to participate in apprenticeship situations (Munroe & Munroe, 1971; Munroe, Munroe, & Whiting, 1985; Parke & Bhavnagri, 1989; Whiting, 1980). Lave and Wenger (1991) argue for the legitimacy of "peripheral participation" when opportunities are structured and made available for children to observe the practices of more capable community members. It appears that these less direct social interactions but adult-directed activities foster knowledge and cognitive skills specific to the culturally sanctioned practices of the community.

Stimulus Attributes of the Task. All cognitive skills are assessed through some mode of representation (e.g., visual, spatial, and auditory). Many studies have examined how familiarity with these cultural tools and materials influence the efficacy with which cognitive skills are expressed. For example, Lantz (1979) found that Indian children showed better classification skills when the stimulus attributes of the task consisted of grains and seeds than when the same task used an array of colors. Similarly, Serpell (1979) compared the effects of media on performance on a pattern production task among children from Zambia and England. Zambian children performed better when reproducing the patterns in a familiar medium (modeling with strips of wire) than when they were asked to reproduce the patterns with a paper and pencil measure. In contrast, the English children performed better with the paper and pencil measure than with the wire medium. When a medium familiar to both groups (clay) was used, however, both groups performed equally well on the pattern reproduction task. Uttal and Wellman (1989) reported how the preschoolers' exposure to acquisition of spatial knowledge through maps may have accounted for their relative competence in subsequent map reading skills.

Communicative Conventions and Courtesies. A prerequisite for engaging in cognitively complex tasks may be related to the skill in the conventions within a community for acquiring and communicating knowledge and skills. An important question is whether children have the skill with certain communicative conventions for demonstrating the cognitive skills they already possess. Conversely, unfamiliarity with particular patterns of discourse may account for less than optimal performance on cognitive tasks. A number of researchers have found that variations in the mode of discourse accounted for differential performance on cognitive tasks (Gauvain & Rogoff, 1989; Kearins, 1981; Lancy & Strathern, 1981, 1989; Miller-Jones, 1989; Siegel, 1991).

Values and Beliefs of a Culture. Although, to our knowledge, there are no studies of human intelligence that have explicitly examined the values and beliefs of a culture as independent variables on cognition, these constructs are embedded in the very tasks and social interactions in which children engage and, consequently, serve an important socializing function in shaping intellectual development. During the past 20 years, Goodnow has commented on these constructs quite extensively. In accounting for the nonindependence of cognition from its contents and context in early cross-cultural studies, Goodnow (1976) argued that such findings reflect the goals and the values of a culture. According to her, cognitive problems or tasks neither exist in a vacuum nor are they ever connected to some abstract set of principles or framework. Rather, they are bounded by a culture's definition of the problem to be solved and its definition of "proper" methods of solution. Goodnow (1990) contends that cultural values contain tacit understandings of what constitutes an appropriate goal and proposes that individuals learn "cognitive values." In other words, culture defines not only what its members should think or learn but also what they should ignore or treat as irrelevant, aspects that she terms "acceptable ignorance or incompetence."

Conclusion

The research reviewed provides good empirical support for the influence of culture on human cognition. It would appear that within any given culture there are certain ecologies or contexts wherein an individual engages in experiences that influence the way cognitions are elicited and deployed for a given task. How much does the inevitable constraint that biology imposes on cognition affect the efficiency and accuracy of the operations of cognitions in such situations? Does the biological constraint channel cognitions along certain predetermined developmental tracks irrespective of the quality or quantity of cultural influences? How susceptible are these cognitions to modification given that the range for the development of cognitive potentials is limited by biological constraints? These are but a few of the questions that ultimately must be answered about human cognition but for which the literature reviewed was largely silent. In Chapter 4, we examine the influences of both biology and culture on the development of cognitions underlying human behavior in tasks defined as "intellectual."

4

Toward a Biocultural Perspective of Intellectual Development

The construct of intelligence has remained one of the most enduring and controversial topics in the history of psychology. Today, we still have not put to rest the genesis question of intelligence that was put forth more than a 100 years ago when Galton (1883), Darwin's cousin, published his heredity thesis about the construct. Jensen's (1969) *The Differences Are Real* and Herrnstein and Murray's (1994) *The Bell Curve* make similar claims that the evidence is substantial for the observed differences in behavior as measured by standardized tests of intelligence. In a different but related vein, other researchers point to the impressive evidence of neural efficiency in accounting for individual difference in intelligence, thus further bolstering the biological argument. There are other alternative explanations, however. Proponents of the cultural view also provide compelling evidence that suggests that observed behavior is not independent of the cultural forces that shape, support, and guide its development and organization. Indeed, the situatedness of human cognition perspective is in direct contradiction to the notion that intelligence is essentially a construct located within the individual. Despite the strong claims on each side of this "either-or" and "how much" debate, it is more likely that an interactionist perspective may shed light on the apparent causal ambiguity surrounding the observed differences in behavior on intellectual tasks.

In this chapter, we examine in detail the major components of an interactionist perspective of intelligence. We begin with a definition of intelligence that is congruent with a biocultural concept. Next, we put forth the underlying assumptions about intelligence and specify what we consider are the key variables for understanding its development and expression. During the course of the discussion, we show how the interdependence of biological potentials with cultural experiences nested within particular cultural niches render attributions for individual differences in intelligence, in either biological or cultural terms, untenable. More specifically, we examine the mechanisms by which cultural experiences within particular cultural niches over time transform biological potentials into developed cognitions. It is these developed cognitions, honed and socialized by cultural experiences, that we believe account for the differences in behavior that are observed and sometimes measured with standardized tests of intelligence.

Definition of Intelligence

Intelligence is a culturally derived abstraction that members of any given society coin to make sense of observed differences in performance of individuals within and between social groups. This notion is similar to the one put forth by Neisser (1976) when he described it as a cultural contrivance created by a people to define what they value as a culture. In the previous chapters, we have come to the realization that the search for an objective definition with universal consensus is a futile endeavor. Horn (1991b) stated it best when he said,

> Efforts to define intellectual capabilities "once and for all" are doomed to failure because not only is the universe of these capabilities so vast that its boundaries are beyond comprehension, but also because it is constantly evolving into a new vastness. (p. 198)

Therefore, we too, like Neisser, have engaged in cultural inventing to define intelligence as the deployment of culturally dependent cognitions in adaptation to meaningful encounters in our environment in a purposive manner. Its expression as behavior reflects the gradual transformation of biologically programmed cognitive potentials into developed cognitions through a process of cultural socialization.

The Biocultural Perspective

The converging evidence from various disciplines that human cognition is context specific in addition to evidence of the strong influence of culture on cognitive development provide the empirical basis for the thesis that intelligence is a culturally dependent construct. More specifically, the evidence suggests that the mind functions and develops within cultural niches and as such the comingling of biological and cultural processes is inevitable. Although the range of cognitive potentials may be constrained by biological programming, which potentials become developed and expressed are under the control of cultural experiences. In this sense, the mental life of individuals is inseparable from the culture that gives it direction, regulation, and meaning; hence, intellectual behavior is more appropriately described as a biocultural phenomenon. From this perspective, individual differences in intelligence are best understood within a developmental framework wherein cultural forces shape the development of biologically programmed cognitive potentials along different pathways toward different end states.

Assumptions

There are four assumptions underlying the biocultural perspective: (a) The interactions between biologically derived cognitive potentials and forces operating within the child's culture are reciprocal, (b) the interdependence of knowledge and cognitive processing in the development of cognition, (c) instruction is a precursor to the development of cognition, and (d) motivation as energy activated from both within and outside the person. A brief explanation of each assumption follows.

Reciprocity

The biocultural perspective asserts that the characteristics of the individual and characteristics of specific characteristics within the child's culture are reciprocally interactive. Biologically derived potentials and other psychologically relevant characteristics are developed and shaped by culture, which itself undergoes change by developing cognitions. All human beings are born with capabilities that enable them to think in complex ways, such as the capacity to encode, scan, transform, reason, store, and retrieve information from memory. Selected attributes within the culture, however, determine when, how, and under what conditions these potentials develop and are manifested in behavior. Similarly, the nature and quality of social interactions and other kinds of cultural stimulation determine how well we organize our thinking and adapt to the ecologies in which we live and grow. The influence is therefore reciprocal or synergistic in that the interplay within and between biological and cultural characteristics results in changes that become the basis for greater and progressively more complex changes in both domains. For a more comprehensive discussion of the interactionist perspective on human development, see Bronfenbrenner (1989, 1993), Ceci (1990), Gordon (in press); Gordon and Terrell (1981), and Lewin (1935).

This process of change, with its reciprocal and synergistic effects, continues over the course of human development. At any point along the developmental continuum, the nature and quality of emerging cognitions are indivisible products of the dynamic weaving of biology and culture. Whether these emerging cognitions will reach their fullest possible expression, remain undeveloped, or show stunted or uneven development depends on two factors: (a) the opportunities and constraints within the culture that may foster or impede their growth and development and (b) the receptiveness or vulnerability of the organism at critical points in time toward these liberating and inhibiting forces operating with the culture. Thus, the number, type, and level of cognitions developed through this process of cultural socialization will consequently vary depending on the confluence of motivational, emotional, social, and cognitive forces operating within both the child's immediate ecologies and the larger culture.

Interdependence of Knowledge and Cognitive Processing

The biocultural perspective assumes that both knowledge and processing play important roles in cognitive development. Consequently, differences in observed behavior as measured by standardized tests of intelligence may be accounted for by differences in efficiency of cognitive processing as well as by differences in the nature and structure of one's knowledge. Empirical studies to date have not established the primacy of one over the other. Some scholars claim that the degree of elaborateness and differentiation of knowledge structures enable individuals to represent the knowledge in memory that makes possible the recognition of new relations and, consequently, the use of existing cognitive operations (Case, 1985; Ceci, 1990; Chi, 1978; Keil, 1984). Other scholars contend that it is the existence of the cognitive operation (Hunt, 1978; Jensen, 1980; Sternberg, 1977a, 1986) or the mental structures (Piaget, 1952) in the first place that enable the acquisition of knowledge. To our knowledge, the "chicken and the egg" question of which one is primary in its influence on cognitive development has not been settled. Nonetheless, it would appear that both the degree of elaboration and the differentiation of knowledge as well as the efficiency of cognitive processing play mutually supportive and complementary roles in the development of cognition.

Instruction and Cognition

Instruction as a form of transactional activity is another tenet of the biocultural perspective. It assumes that the development of cognition is facilitated when it becomes the object of instruction. It may be direct, as in significant other-child dyadic relationships, or it may involve less direct social processes, as in apprenticeships situations in which semistructured opportunities are created for observational learning by the child of expert practices in a given domain of interest. It is based on a Vygotskian notion that instruction is effective when it is directed at those cognitive functions not yet completely formed that lie in the zone of proximal development. The role of the instructor is to facilitate the emergence or development of these nascent cognitions.

Internal and External Motivation

The biocultural perspective holds that there are two aspects of motivation in the development and expression of cognition. The first describes the capacity of the individual to arouse attention and interest in environmental stimuli, to sustain the intensity of effort, and to direct one's energies toward the completion of a task without external feedback or reward. The second describes the mechanism within specific environmental stimuli that activates or triggers initial interest and attention in a task, which also serve to sustain intensity of effort in tasks and to direct one's energies toward fulfillment of goals. Banks and colleagues' (1979) conception of the embeddedness of relevance and interest in tasks, Gordon's (1991) prompting force within environmental stimuli, and Ceci's (1990) context elicitors all convey the notion that motivational forces also lie within the culture.

These are our working assumptions as we seek to develop a biocultural theory of intelligence. It is developmental and reflects a strong interactionist position of the dual role of biology and culture in accounting for its development and expression. We express the reciprocal interaction with the following equation:

$$D_t \rightleftharpoons B_t = f_{(t-p)}\mathrm{ST}(\mathrm{PE})_{(t-p)}$$

where B represents intelligence as cognitive behavior and t is the particular point at which it is observed; the symbol f stands for function and is used as an indicator for developmental processes through which conditions and attributes of the person and the environment are reciprocally interactive in ways that produce continuous change in cognition overtime; t – p is the prior period when the reciprocal interactions between person and environment conditions and attributes were occurring to produce the cognitive behavior observed at the particular point in time of observation; ST is the sustaining and threatening forces or conditions; PE is the person and environmental factors; D is cognition as a developmental outcome observed at a particular point in time; and the symbol $\rightleftharpoons$ is used to indicate that B_t and D_t can be used interchangeably.

Of course, representation of the interactionist perspective in equation form is not a new idea. More than 30 years ago, in his seminal study of the psychology of human behavior, Lewin (1935) advanced the notion that behavior is a joint function of person and environment: $B = f(\mathrm{PE})$. Since then, other scholars have transformed this classic formula with substitutions. For example, Gordon's (1977) conception of sustaining and threatening forces as well as existential and objective realities of the person in the person-environment interactions led to a reformulation: $B = f[\mathrm{o(SPE)o(TPE)e(SPE)e(TPE)}]$. Bronfenbrenner (1993), couching person-environment interactions in developmental terms, argued that development is a joint function of person and environment ($D_t = f_{(t-p)}\,(\mathrm{PE})_{(t-p)}$. We have blended various aspects of the work of these scholars in the way we have represented the developmental nature of the reciprocity of person-environment interactions and its expression as behavior at a particular point in time. In the sections that follow, we examine the various components of our emerging biocultural theory of intelligence in greater detail.

Characteristics of the Developing Child

Cognitive functioning, although important, is indicative of but one aspect of a child's life. Other psychological processes—social, emotional, and motivational—play an important role as well. In our biocultural perspective, we begin with the position that differences in intelligence are more meaningfully understood within a developmental frame and as such consider other characteristics of the child as well. The section that follows begins with an examination of the concept of biologically derived cognitive potentials and culturally dependent cognitions. In addition, other psychologically relevant characteristics are considered that may have implications for the development of the human cognition that some of us regard as "intelligence."

Biologically Programmed Cognitive Potentials

Despite the seemingly irresolution of the genesis question of intelligence, it is difficult to dismiss the possibility that some basic cognitive functions as well as the dynamic organization and structure of the brain create a biological predisposition for human development. These biologically derived cognitive potentials include mental operations for attending, encoding, and scanning environmental stimuli as well as operations for transforming, storing, and retrieving environmental input. A speed factor appears to be pervasive within and across these cognitive operations. (See Ceci, 1990, for discussions of these natural cognitive functions.) We share Weinberg's (1989) view that our genes set limits in terms of the range of possible reactions for these labile cognitive potentials. The nature and quality of environmental encounters in a given culture, however, determine whether or not the full range of gene reactivity is developed and ultimately expressed.

Culturally Dependent Cognitions

We take the Vygoskian position that higher classes of cognitive function (e.g., reasoning, logical memory, planning, fluency in language, speed of decision making and retrieval of information from memory, etc.) are higher human functions conceptually distinct from the lower classes alluded to earlier. According to Vygotsky (1978), lower (natural) classes of cognitive functions are reorganized and transformed according to the means and social goals established by a culture. The transformation is made through social mediation, symbolic tools, and materials, all of which have psychological salience for the developing child. (For a more comprehensive discussion of these issues, see Newman & Holzman, 1993; Wertsch, 1985.)

We label these higher cognitive functions culturally dependent cognitions because it appears that it is through a developmental process of cultural socialization that account for their restructuring and transformation from lower cognitive functions (biologically constrained potentials). At any point in the developmental history of a child, through task analysis it should be possible to determine the nature and quality of these developed, culturally dependent cognitions. To the extent to which we need to use them to discriminate among individuals or groups, then a variety of quantitative and qualitative procedures may be used. As we argue later, differences in the expression of these cognitions may be more of a function of the nature and quality of learning experiences within certain cultural niches to which the child has been exposed rather than to defective genes or faulty neural processing. We make no claims for the number and level of thinking involved in these cognitions as other researchers using factor analytic and information processing models have done. This is not an attempt to dodge questions of generality versus specificity and one versus many regarding the nature of intelligence. Indeed, to date the status of our technology and statistical techniques are unable to address the causality of intelligence issue without polemical debate. As such, these questions have deliberately not been the focus of our attention.

Other Psychologically Relevant Characteristics

Demographics

Demographic characteristics of the person may or may not have implications for the developing child. In multicultural societies such as the United States, demographic characteristics, such as age, race, class, gender, and ethnicity, are not merely static variables used to conveniently distinguish different categories of people. Rather, for many individuals, they are psychologically charged constructs with implications for how individuals react to situations, events, or people in the environment as well as how others perceive and react to them. Phinney (1996) provided a discussion of the psychological importance of ethnicity in U.S. culture, but we think other demographic characteristics may have similar significance for some individuals as well. When these characteristics are considered in relation to cognitive development and more specifically intellectual functioning, they take on special meaning and relevance for individuals, particularly those who identify with a cultural group within a dominant social order. The ideological orientation, opportunity structures, and patterns of social, political, and economic interchange that are embedded within the dominant culture determine to a significant extent the course and conditions of cognitive development. There is reason to expect that the process of cognitive development may operate differently for ethnic and racial minorities as well as for males and females. To the extent that this is the case, any consideration of intellectual functioning must include its interactive influences with those aspects of demographics that have psychological relevance and meaning for some individuals and groups. (The interested reader is referred to Boykin, 1983, for a discussion of the psychological functioning of some racial and ethnic minority groups within the U.S. society.)

Response Tendencies

Our review of the literature suggests that there are a variety of cognitive, emotional, and cultural patterns of an individual's response to specific environmental stimuli—situations, persons, or events. To describe these idiosyncratic responses to situations, a variety of terms have been used: affective response tendency or temperamental style (Gordon, 1988; Thomas & Chess, 1977), cognitive style (Messick, 1976; Shade, 1982; Shipman & Shipman, 1985), learning style (Dunn & Dunn, 1978), cultural and behavioral style (Boykin, 1979; Hale, 1982; Hilliard, 1976), developmentally instigative characteristics (Bronfenbrenner, 1993), and cognitive response tendencies (Gordon, 1988, 1991). In reviewing this work, we were unable to find unequivocal findings for the relationship between these stylistic modes and behavior. Some of them appear to be dynamic dispositional traits that show a high degree of stability that may be minimally responsive to environmental stimulation. Others, however, seem quite labile and consequently are responsive to situational demands. Some of these latter "response tendencies," as Gordon (1988, 1991) labels them, are motivational and cognitive in nature in that the child shows initial interest and attention to certain types of tasks and remains engaged over a prolonged period of time until task completion. When these personologic qualities are considered in relation to cognitive development, and more specifically cognitive functioning, they are likely to interact with biologically derived potentials as well as with other psychologically relevant demographic characteristics of the child in nontrivial ways.

In summary, a developmental perspective allows a conception of the child as a functional whole wherein cognitive functioning does not operate in isolation of other psychological characteristics of the child. So as not to run the risk of overgeneralization or distortion regarding observed differences in behavior measured by intelligence tests, we consider other aspects of the child in our biocultural perspective of intelligence. Our reading of the literature suggests that these characteristics of the child are quite dynamically interactive and are likely to influence the course and outcome of subsequent cognitive development. Such characteristics, however, do not exist in isolation of the cultural environment in which development unfolds. Rather, they interact with certain characteristics of the cultural environment in ways that substantially affect the development of biologically derived cognitive potentials. The product of these interactions that we observe as behavior is consequently biocultural in nature. To better understand the psychological effects of culture, we turn now to a discussion of its defining attributes and the mechanisms by which it enables the transformation of biologically derived cognitive potentials into what we call *culturally dependent cognitions.*

Culture

The picture that emerges from both the theoretical and empirical research on culture, cognition, and behavior is one that suggests that culture permeates the daily life of a people and as such plays a pivotal role in human development. In terms of its location in cognitive development, and more specifically intelligent behavior, it seems to us that it has the potency to shape, direct, and transform biologically constrained potentials into developed cognitions. Over time, these developed cognitions, honed by the process of cultural socialization, are reflected in those special capabilities that members of a culture use to meet the demands of their social, economic, and technological environment.

Our position has been inspired by Vygotsky's (1978) sociocultural theory and thus many of our ideas are consistent with those of other researchers who argue that thinking does not exist outside of the activities in which people engage and the cultural practices that support and maintain desired patterns of cognitive development (Boykin & Allen, 1991; Ceci, 1990; Cole, 1988; Rogoff, 1990; Wertsch, 1985). From this work, both empirical and theoretical, we were able to discern at least three broad defining attributes of culture with psychological significance for the cognitive development of the child:

(a) belief system, (b) symbol system, and (c) language system. Although there is conceptual overlap among these systems, each one provides essential information regarding the relationship between cognition and behavior. A brief description of each attribute follows.

Belief System

Beliefs are interrelated concepts that govern the day to day lives of a social group. These include norms that describe the social standards and expectations for behaviors that, according to Berry (1976), people regard as right, proper, and natural. Beliefs are often implicitly understood and reflect a tacit consensus of assumptions about individuals and groups and their place within the society. Beliefs reflect a "mind-set" that remains deeply entrenched in the psyche of a people despite the passage of time. The term is sometimes used interchangeably with the notion of "worldview" or an "ethos" that according to Mbiti (1970) includes concepts such as understanding, attitude of mind, and perceptions that influence the way people think, act, and speak in various situations of life. Goals define the targets and expectations and give focus for people's energies and thought. Collectively, we refer to these interrelated concepts as a value system because it is likely that these concepts have their greatest impact as a cluster rather than a single entity. With respect to cognition and behavior, these intangible yet powerful attributes of a culture not only establish the opportunities and constraints for the types of cognitions about which judgments are made but also provide structure, direction, and regulation for its development. The daily activities in which individuals engage and the practices that support and maintain these values function as essential resources fueling their motivations and their thinking along particular pathways toward particular ends.

We find support for these ideas in Vygotsky's (1978) *Mind in Society*. In that work, he argued that at any given point in its history, a culture both defines and sets limits on options for individual development and it is to the flow of history that we look for evidence of psychological development across all domains—cognitive, emotional, and social. This means that the characteristics of the person as described earlier may be encouraged or discouraged depending on the social, economic forces operating within a culture at any given point in time and the structured relationships embedded within its institutions. More specifically, the nature and quality of cognitions are often indirectly shaped by the belief system that is implicitly reflected in these institutions.

Brofenbrenner (1979, 1993) makes a similar point about the overarching influences of the macrosystem in his discussion of the role of ecology in cognitive development. Culture at this macrolevel, although critical, does not tell the full story of its impact on the developing child. For its influence at a more direct level—in face-to-face adult-child or capable peer collaborations—we turn to a discussion of culture's other defining attributes.

Symbol System

The symbol system describes the technologies (e.g., linguistic, pictorial, numerical, and gestural) that enable the development and ultimately the expression of cognition within any given culture. It is through the use of a symbol system that the child acquires knowledge, the differentiation and elaboration of which enable him or her to make connections to events, objects, and persons within his or her environment. Some cultures use more than one symbol system, so it is not uncommon that task demand reflect different permutations in content representation. For example, a task may require the processing of knowledge acquired through a dual modality of content: verbal-auditory, visual-spatialization, auditory-spatialization, pictorial-auditory, and kinesthetic-auditory. Of course, even more complex permutations of these dual modalities may be represented in some tasks. We speculate that the level of cognitive complexity of some tasks is to some extent a function of the sophistication in which the task content is represented. In principle, therefore, we make the case that the efficacy in which children are able to do Spearman-like tasks of "eduction of relations" and the "eduction of correlates" depends to a large extent on children's familiarity with the symbol system in which such reasoning tasks are represented. In cultural terms, we contend that the level of generality at which cognitions are expressed in behavior is dependent on the opportunities a given culture affords its members to access and use its symbol system(s) in meaningful tasks.

Again, we look to Vygotsky's (1978) conception of the symbolic feature of tool-mediated activity to better understand the mechanisms by which lower cognitive functions are transformed into higher cognitive functions. According to Vygotsky, the type of material tools places corresponding demands on one's human mental faculties.

They are not merely a collection of individual implements. They take on psychological significance when used to represent phenomena in collective human interactions. In Vygotsky's seminal work (1978), he identified some ancient tools that served psychological functions such as "tying knots" and "counting fingers." Tying knots was used as a mnemonic tool to facilitate retrieval of information from memory. Using fingers to count functioned as a support in higher cognitive processing that involved basic arithmetic operations. He also mentioned other more advanced symbolic mediators that included artificial and natural languages.

Language System

A language system describes a number of different ways a culture systematically communicates ideas, feelings, and thoughts through the use of words, sounds, gestures, or signals with commonly understood meanings. It shares with the symbol system the modality for communicating environmental stimuli. Unlike the symbol system, however, the emphasis here is on the sociolinguistic conventions of a cultural group for organizing social interaction between an adult-child or peer collaborations or both. These culturally valued media act as a vehicle for dynamic and mutual engagement of cognitive tasks and through which the child's knowledge and thinking undergo organization, restructuring, and transformation. Over time, through these reciprocal processes between adult-child or capable peer collaboration, the child develops a system of knowledge structures and cognitive skills that are congruent with the symbol and belief systems of his or her cultural group. These culturally coded knowledge structures are then used as interpretative frameworks for using cognitions in the acquisition of new knowledge. Although the precise relationship between cognition and knowledge remains unclear, we concur with the positions of Keil (1981, 1984), Chi (1978), and Ceci (1990) that the degree of elaboration and differentiation in the representation is perhaps the mechanism that makes possible the recognition of new relations and consequently the use of existing cognitive processes. Also, it is from this perspective that we speculate that the efficiency or automaticity with which some children respond to intellectual tasks both in and outside of school may be a reflection of their well-elaborated and differentiated knowledge structures and developed cognition honed through this reciprocal process of the language system. Conversely, it is quite likely that the difficulties that children from linguistically diverse backgrounds experience on standardized tests of intelligence may have more to do with incompatibility between the sociolinguistics patterns of the child and the tester than with innate capacity to think abstractly or to retrieve information quickly. When we consider that the belief and symbol systems of these children may be different from those of the culture that supports the construction and administration of standardized tests of intelligence, their plight is multiplied exponentially.

Once more, we turn to Vygotsky and colleagues for the claims we make about the psychological salience of a language system. Earlier, we alluded to the symbolic aspects of tool-mediated activity. The other dimension of Vygotsky's mediated activity thesis involved another individual or capable peer. According to Vygotsky (1978, p. 57), "Every function in the child's cultural development appears twice: first on the social level, and later on the individual level; first between people (*interpsychological*)[italics added] and then inside the child (*intrapsychological*) [italics added]." He claimed that social mediation involves four concepts that account for its role in shaping cognitive development. First, knowledge in one's culture is socially transmitted by adults and capable peers to children. Second, joint participation in the range of activities determined by the culture allows for certain cognitive skills to be practiced and demonstrated by adults so that the children's current cognitive functioning may be modified or strengthened. In this way, the adult helps to shape the child's existing skills to better suit the demands of the culture. Third, new cognitions are cultivated when the adult or capable peer shares in the responsibility for the task with the child. Assuming the role of an expert tutor, the adult models, corrects, clarifies, and explains concepts to the child so that the child attempts and completes the task according to the criteria established by the culture. Finally, the independent use of new cognitive abilities is encouraged when the adult or more capable peer works with the child on cognitively challenging tasks that the child could not have successfully completed without guidance and support. Working with the child in their *zone of proximal development*, the adult models the task's appropriate behaviors, directs the child's attention to alternative procedures or approaches to the task, and encourages the child to try out his or her embryonic skills on some portion of the task. As the child's ability develops, the adult gradually reduces instructional support and allows the child to assume greater independence in task solution. It is this type of social scaffolding that Vygotsky suggested as the mechanism for change in cognitive development.

In summary, we think that each cultural system has psychological significance for the developing child. We speculate, however, that in any given culture, these systems interlock and the collective impact on the child is greater than is any one operating in isolation. How, then, do these influences coalesce in ways that affect the cognitive functioning of the child? We use two metaphors, "learning experience" and "cultural niche," as a way to operationally seek an understanding of the psychological dynamics operating within and across the three cultural systems.

Learning Experience

For us, the concept of learning experience comprises a cluster of social, emotional, motivational, and cognitive ingredients for enabling the transformation of biologically constrained cognitive potentials into culturally dependent cognitions. It embodies the defining attributes of culture in ways that the boundaries for the three symbol systems are virtually seamless. More specifically, every experience of the developing child has at least three features: persons (significant other(s) and the child), the tasks (symbolic mode of representation and level of complexity and motivational properties), and processes (social interaction). In practice, either tasks or processes may be subjected to manipulation as a way to better understand its unique effects on development or behavior of the person(s) under investigation. These localized treatments and influences, however, should not obscure the reciprocal interrelationships among these three variables and their impact on the developing child. In the following sections, we consider the findings from research that we think represent attempts to study selective aspects of experience.

Person-Process

Social interactions describe the dynamic processes of mutual engagement that occur between adults, capable peers, or any significant others and the child. They instantiate the language system of a culture by which the significant other, through mediation, structures the tasks in ways that encourage the child to focus on his or her thinking as he or she tackles the task.

The literature does provide some evidence for the importance of adult-child and peer collaboration in tasks requiring the demonstration of developed cognitions. For instance, Artzt and Armour-Thomas (1992) found that variation in mathematical problem solving was associated with the differences in the nature and quality of verbal interaction among seventh-grade students working in small groups. In that study, the ratio of metacognitive to cognitive behaviors was higher for groups that showed higher levels of peer collaborations. Other investigators have documented the gradual evolution of cognition when adults organize the learning environments of children using the principle of social scaffolding. For instance, in the studies conducted by Palinscar and Brown (1984) and Wertsch (1979), the adult was able, through different levels of interaction over time, to socialize the development of self-regulatory skills in children. Finally, Mackie (1980, 1983) studied the effects of social interaction on performance on spatial reasoning tasks among children from European descent and children from Maori and Pacific Island descent in New Zealand. The findings revealed that the children from Maori and Pacific Island descent were more passive in their interactions, especially when paired with more capable peers as partners, and their performance was lower than that of their counterparts from European descent. Mackie hypothesized that differences in the values regarding social interaction in the two cultural groups may have contributed to the differences in effort and performance.

Person-Tasks

Symbolic Representation of Information. Modes of representation are instantiations of a symbol system of a cultural group and include manipulatives, computers, maps, charts, various types of musical instruments, written script, and forms of counting. Acquisition and mastery of knowledge as well as the efficiency and accuracy of cognitions depend in part on the mode of representation of the task demands.

A growing body of cross-cultural research has reported differential effects of various symbol systems on children's cognitive performance. The underlying assumption of these studies suggests that the nature and quality of an individual's experience with the mode in which the task is represented will positively influence his or her performance in that task. For example, Stigler (1984) examined the use of the abacus as a tool for mathematical operations and mental calculations of Japanese students and found that intermediate and expert abacus users used a "mental abacus" when they performed mental calculations. It appears that a mental representation of problems on an abacus enhances remembering specific skills. In another investigation, Lave (1977) compared the arithmetic skills of Liberian tailors with school and tailoring experience using a format used in school with one used in tailoring. Findings indicated that experiences with schooling and with tailoring were related to solving arithmetic problems with the respective formats.

Motivational Stimuli. An important characteristic of a task that is likely to influence performance is the extent to which it has properties that are likely to attract and sustain attention and emotional investment until its completion. Again, we look to the culture and ask the following questions:

- What kinds of activities arouse individuals' attention and interest?
- What kinds of activities encourage them to sustain that effort with a level of intensity until task completion?
- What kinds of cultural practices support, maintain, and validate a high level of energy expenditure?

To the extent that the individual sees relevance and value for himself or herself or that the goal is worth pursuing, then positive outcomes can be expected.

To our knowledge, systematic study of this issue of the linkage between cultural motivation and developed cognition has remained relatively unexplored. There is some indirect evidence, however, that holds promise in this area. For example, Boykin (1982a) and Tuck (1985) reported that African American children's task performance was markedly better when the task context afforded a higher rather than a lower level of variability, the former being more congruent with the amount of variability present in the home life of these children. In a similar vein, Boykin and Allen (1988) and Boykin (1991) found that low-income African American children's task performance could be enhanced when the learning context afforded a greater rather than a lesser opportunity for movement expressiveness. Collectively, these studies suggest that the relatively high levels of sensate stimulation afforded in many African American home environments play a nontrivial role in observed task engagement and persistent behaviors. Similarly, motivational dynamics may have accounted for the efficacy in mathematical problem solving in Carraher, Carraher, and Schliemann's (1985) and Saxe's (1991) work with Brazilian street children. For example, the candy selling observed by Saxe was tied to the primary goal of the children—making money—and it is this motive that may have accounted for the speed and accuracy of their mathematical calculations.

Level of Cognitive Complexity. Level of complexity refers to the cognitive difficulty of the task—both its form of representation and its type and level of cognitive processing. Our reading of the empirical research suggests that researchers have different conceptions of this aspect of task demand that are correspondingly reflected in the analysis and interpretation of their findings. Some studies examined the level and type of cognitive processing in tasks involving reasoning (Pelligrino & Glaser, 1979; Sternberg; 1977), comprehension (Sternberg & Powell, 1983), and metacognition (Brown, 1978; Sternberg, 1986). Findings were usually interpreted in terms of differences in the efficiency of processing or use of cognitive strategies. Other studies examined the nature and quality of knowledge demands of the task (Ceci, 1990; Chi, 1978; Keil, 1984). Findings were often interpreted in terms of differences in knowledge representation. It may well be that both perspectives and interpretations are correct.

Knowledge and process appear to play different but complementary roles in cognitive development and, as such, behavior at any point in the child's development is likely to reflect the effects of a reciprocal interaction of knowledge representation and cognitive processing.

Contexts as Cultural Niches

We use the term *niche* in an effort to locate those particular ecologies within a culture that provide psychologically meaningful experiences for the developing child that are likely to nurture his or her development. It is a concept borrowed from biology—the "ecological niche"—used to describe the relationships between an organism and the environments in which it lives and grows. The term has also been used in developmental psychology. For example, Gauvain (1995) and Super and Harkness (1986) used it as a way of examining the simultaneous psychological and cultural influences on human development. Similarly, Bronfenbrenner (1979, 1993) used it to consider the multiple contexts wherein interactions of a changing organism in a changing environment unfold over the course of cognitive development. In an analogous manner, we think that experiences, as they relate to the development of biologically derived potentials, are embedded in settings in which person-process-task interactions occur. These settings must involve face-to-face relationships between the significant others and the child in which opportunities are provided on a consistent basis for mediating the child's engagement with the task demands. Feuerstein and colleagues (Feuerstein, 1990; Feuerstein, Rand, & Hoffman, 1979; Feuerstein, Rand, Hoffman, & Miller, 1980), building on Vygotsky's work, contend that it is at this face-to-face level where mediated learning takes place and constitutes what they call the proximal determinant on cognitive development. Other factors in the culture may influence cognitive development, but the effects of these distal determinants occur mainly through their influence on mediated learning experiences between the significant other and the child. Bronfenbrenner (1979, 1993) makes a similar argument in his discussion of the role of context in cognitive development.

The cultural niches wherein critical developmental processes between significant other and child occur include the home, the school, the peer group, and the community. Of course, other combinations of cultural niches can serve similar function. For example, for some adolescents, the peer group, the community, and the church may provide experiences that are likely to foster the development of biologically constrained potentials along certain trajectories. For other adolescents, the school, the home, the church, and the peer group through its learning experiences may foster the development of the same biologically constrained cognitive potentials but along different trajectories toward different ends. The point is that regardless of the type of cultural niche, each one is construed as a social-psychological nexus that, through the experiences made available for the child, offers structure, direction, and regulation for the development of biologically constrained potentials. Needless to say, not every culturally dependent cognition that emerges through this process of cultural socialization in every niche will be equally valued by the larger culture in which it is nested. Indeed, it is quite plausible that some cultural niches may have their own belief, language, and symbol systems that are different from those of other cultural niches as well as the larger culture. In multicultural societies such as the United States in which many cultural groups coexist, one can only wonder at the tremendous variations in culturally dependent cognitions that must have developed both within and across cultural groups. Figure 4.1 illustrates the types and level of relations within and across cultural niches as well as the social context of the larger culture within which these niches are embedded.

Conclusion

We have tried to make the case that observed behaviors at any point in a child's development are more precisely defined as culturally dependent cognitions. As members of the species *Homo sapiens,* we are born into this world with certain dynamic characteristics including cognitive potentials that create a biological predisposition for human development. How these characteristics develop, along what pathways, and toward what end states, however, are determined by the reciprocal interplay of these characteristics with equally dynamic experiences within particular zones of the culture wherein the child develops and functions. From this biocultural perspective, differ-

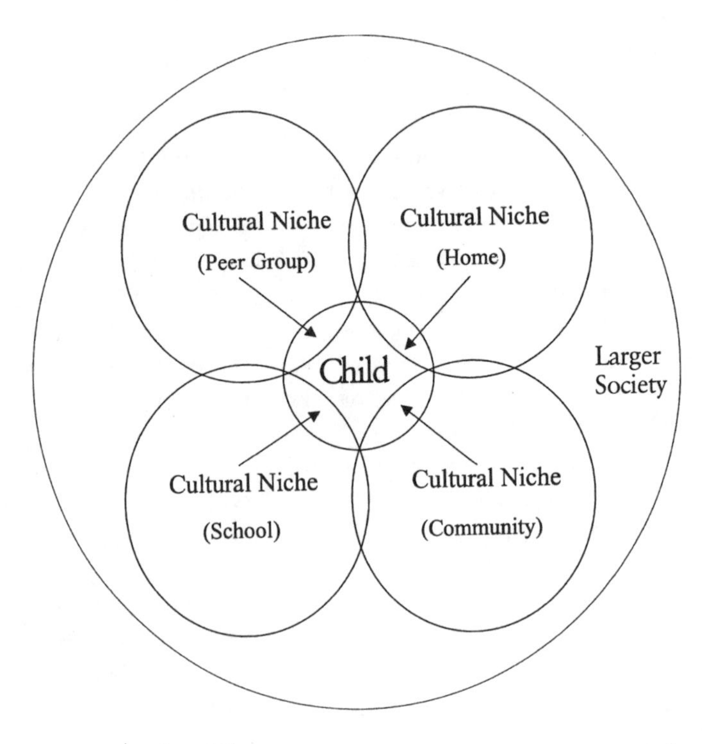

Figure 4.1. Biocultural System

ences in behavior that some of us label as intelligent reflect the extent to which the culture has differentially shaped the development of those biologically constrained cognitive potentials under investigation. These issues of person-process-context interactions, to our knowledge, have been unevenly explored in investigations of biology, culture, and intelligence. This is particularly troubling given the fact that in some cultures, such as the United States, judgments of intelligence have such far-reaching implications for how much and to whom the nation's resources are allocated. We hope that the biocultural perspective offers another lens through which to study the development of intelligence and to analyze, interpret, and draw implications of findings consistent with such a perspective. In Chapter 5, we describe the development of a four-tier assessment system of intelligence consistent with the notions of a biocultural perspective.

PART II

Intellectual Assessment and Culture: A New Paradigm

5

The Evolution of the Biocultural Assessment System

The Early Years

This assessment system has a long history that began in our private practice in 1986. It is based on the assessment of approximately 2,500 children who, over a 10-year period, were referred to Multicultural Educational and Psychological Services in Hempstead, New York. These children were referred from approximately 42 school districts and 150 schools in New York, Connecticut, New Jersey, California, and Washington. Approximately 7% of the children assessed were of African American heritage and 3% were Anglo-Americans. The remaining 90% were bicultural, originally from the United Kingdom, Puerto Rico, the Dominican Republic, Columbia, Peru, Ecuador, Mexico, Panama, Costa Rica, Haiti, Trinidad/Tobago, Barbados, Jamaica, Grenada, Guyana, Antigua/Barbuda, St. Kitts/Nevis, St. Vincent/Grenadines, St. Lucia, Montserrat, and Dominica.

As students in training in the early 1980s, there were some nagging concerns that the manner in which we were being trained to assess intelligence and interpret its findings was incongruent with our cross-cultural experiences given our understanding of intelligent behavior. The interpretation of differences on the IQ measure in terms of biological inferiority or cultural deprivation seemed equally untenable. When we examined the literature and the subjects for whom these judgments were made, they paralleled race, class, and linguistic differences. It was therefore rather disconcerting that the research on intelligence overwhelmingly revealed that minority (African American, Latino, and Native American) children on an average performed as much as one standard deviation (15 or 16-scale score IQ points) below that of their Anglo-American counterparts. In addition, the research also revealed that during a 3-year period children's IQ very rarely changed more than 3 IQ points as assessed by standardized tests of intelligence. We could not come to terms with our growing disenchantment with the assumptions that seemed to underlie the practice and interpretations of findings. We believe that all human beings have the capacity to remember, to reason, and to acquire and use knowledge. The speed and accuracy with which these capacities are deployed are dependent on the kinds of experiences from which individuals are socialized and the value system of the culture that supports the expression of these cognitions. This discomfort about this entire intellectual assessment enterprise led to a life-long commitment to investigate these issues more fully.

Upon graduation in the mid-1980s, we formed an agency—Multicultural Educational & Psychological Services—whose goals were

1. to provide psychological, educational, psychiatric, and speech for ethnic, linguistic, and other culturally diverse families.
2. to conduct research and provide workshops and consultation services to these families as well as to various agencies, school districts, and organizations throughout the tri-state areas.

Psychometric Assessment

When we first began assessing children at the multicultural agency, we adhered strictly to the standardized procedures as we were initially trained to do. Postassessment interviews of the 650 children assessed between 1986 and 1988, however, revealed that standardized measures of intelligence did not provide a complete understanding of the full range of a child's ability. Although cognitions related to memory, reasoning, and knowledge acquisition and application were present on the IQ tests, they still did not capture the breadth of the children's experiences and the multiple contexts in which these experiences were manifested differently. Anecdotes such as "I know this, but I can't describe it" and "Could I tell you what the word means in a different way" led to a theoretical review of the literature on the receptive and expressive vocabulary of the importance of contextually embedded words (Sattler, 1988; Sternberg, 1985). We also examined the literature on knowledge-acquisition processes from Sternberg's triarchic theory of intelligence and the general literature on the availability and accessibility of cognitive strategies from long-term memory. Furthermore, we became familiar with the works of Keil (1984) and Chi (1978) regarding knowledge and its structure in long-term memory. Hence, the idea of contextualizing vocabulary as a means of assessing a child's cognitive potentials was born.

Furthermore, many children, when faced with the arithmetic subtests that required them to solve problems mentally, claimed that "I can't do this in my head." When given the opportunity to use paper and pencil, upon completion of the testing, they were often correct. We found empirical support for this procedure from the research that examined the use of materials and tools to support mathematical understanding and accuracy (Stigler, 1984; Vygotsky, 1978). Hence, the idea of allowing the use of paper and pencil as a form of potential assessment was developed.

Another clinical observation that was evident during those early years was the fact that many children from culturally diverse backgrounds seemed totally lost when given blocks and puzzles to manipulate. On several occasions, some children asked, "What do I do with this?" even after the standardized explanation was provided by the examiner. When a more direct explanation of the task demands was provided and when the children were allowed to go beyond their ceiling points, significant gains were noted. It became clear that they were learning as they went along and that the task of building blocks and puzzles was novel and unfamiliar to these children. Therefore, the familiarity with blocks that was so endemic to the typical American 5-year-old child was foreign to many of these culturally diverse children. The theoretical insights of Vygotsky (1978) regarding the role of social mediation in ascertaining the zone of proximal development in children whose cognitions are not yet well formed were examined. Also, we examined the results of Feuerstein, Rand, Hoffman, and Miller's (1980) Instrumental Enrichment Program and Lidz's (1987) dynamic assessment strategies. Hence, the development of the test-teach retest assessment measure.

In general, an ongoing debate in intelligence testing is that of speed of mental functioning. Carroll (1993), De Avila (1976), Eysenck (1986, 1988), Horn (1991a), Jensen (1979), and Sternberg (1986) noted that assessing culturally different children via timed tests confuses the measurement of ability with measurement of aspiration because little regard is given to children who are not culturally trained to work under timed conditions. Gopaul-McNicol (1993) found that most Caribbean children have difficulty completing tasks under time pressures because this represents the antithesis of what their culture dictates. On the contrary, slow and careful execution of their work is highly valued so that even if the child is aware that he or she is being timed, he or she may ignore the request by the examiner for a quick response and instead execute the work methodically and cautiously. Several children stated in a jocular manner, "Is this a test? I want to get it right." They were more concerned about accuracy than completing the test in a speedy manner. As such, scores tend to be lower for such students on timed tests, which comprise most of the nonverbal subtests and some of the verbal ones. Although speed may be essential in such situations, most of everyday life's events do not require decision making in a few seconds typically allotted for problem solving on IQ tests. The important issue here should not be one of total time spent but rather time distribution across various kinds of processing and planning events. The practical point to be made from this is that students should not be penalized for not completing a task in the allotted time (speed test). Instead, they should be credited for successful completion of the task regardless of how much time it took (power test). As such, the idea of suspending time as a form of potential assessment was born.

The Later Years

Psychometric Potential Assessment

In 1988, we secured permission from parents to experiment on the various ways of best assessing a child's potential by "stepping away" from standardized procedures as mentioned previously. By 1990, after assessing approximately 625 more children, we felt we had mastered a standard procedure for potential assessment. Between 1990 and 1992, we utilized this standard procedure, which we called psychometric potential assessment; after assessing an additional 600 children, however, we found that many parents were disenchanted with our findings, often claiming that "my child knows more," "my child can do this," and "my child is not mentally retarded or deficient."

Ecological Assessment

Parents invited us to visit their homes and observe the children in their natural settings. A significant change in the development of our assessment procedure resulted after these home and community visits. It was clear to us that these children were by far more skilled in all the areas assessed, but for reasons unknown to us at the time, these children were unable to attain success on the IQ tests whether they were assessed to their potential or whether adherence to standardized testing was done. Children were seen building chairs and tables, fixing bicycles and cars, repairing fans, televisions, and other electrical appliances and yet they could not put blocks and puzzles together. The works of Helms's (1989, 1992) cultural and item equivalence and Sternberg's (1986) *Beyond IQ* were examined in 1992 and 1993 as we assessed approximately 350 more children. This led to the development of the third tier of our assessment system—the Ecological Taxonomy (Ceci, 1990).

Other Intelligences Inventory

Also in 1992 and 1993, we observed that many children who were found to be deficient on psychometric or potential psychometric tests were found to be talented in various athletic and musical arenas. One child who was clearly a Down syndrome boy played the violin and the piano with fluency and poise. One young man was such a genius in basketball that his parents encouraged him to pursue this talent. Today, he is a professional player and attributes his career choice to the support he received from his parents after they read the examiner's report. This observation and finding led to our examination of the literature that deals with content in which cognitions are embedded, including the work of Gardner's (1983) multiple intelligences. Out of this emerged the last tier of our assessment system—the Other Intelligences Inventory.

In 1993 and 1994, we assessed approximately 250 more children utilizing this four-tier assessment model, now coined the Biocultural Assessment System. In addition, assessment measures were developed to assist in this assessment process—the Family Assessment Support Questionnaire, the Stage of Acculturation Measure, and the Teacher Questionnaire. Together, these nonpsychometric measures beautifully complimented the psychometric IQ measure to give a more comprehensive picture of a child's cognitive functioning.

In 1994, we began to train graduate students in school psychology programs, and by 1995 several New York state school districts contracted us to conduct trainings with their professional staff psychologists using this comprehensive assessment system with the following goals:

1. Determine a more accurate profile of a youngster's potential for learning and for intervention
2. Help psychologists to understand that many variables contribute to and explain performance and that psychological assessment is both a formal and an informal process that occurs in several contexts—the school, the home, and the community

Length of Time for Training

From our experiences in the training of this model, we found that it takes a professional person who has been in the field for 2 or more years approximately 2 full days of training. A professional in the field with under 2 years requires approximately 4 days of training.

A student in training would need approximately one semester (two or three 1-hour classes per week) after he or she had been trained to use traditional psychometric IQ standardized tests for at least one semester. This is because to appreciate and value the benefits of this biocultural assessment system, some exposure to traditional measures of assessment may be useful.

Why the Need for Such a Large Sample and for Such a Long Time Period?

Because we were attempting to change the manner in which IQ testing is conducted, we tried to secure ecological validity by building on a large sample size because experts in the assessment of minority children (Cummins, 1991; Hamayan & Damico, 1991; Hilliard, 1996; Samuda, 1975) espouse the notion that a more comprehensive understanding of within-group differences of minority children is needed before intergroup comparisons can be made. These researchers emphasize that intergroup comparisons foster the view that minority children are abnormal, incompetent, and underdeveloped. They recommended that a thoughtful analysis of the role of situational, ecological, cultural, and systemic factors that shape the behavior of minority children need to be understood by studying these children in their own right. Of course, this position weakens the view that a control group of Anglo-American children is needed for adequate interpretation of the research findings of African American or other non-white children. Herein lies the rationale for studying such a large sample of children. The idea was to understand, in this case, a particular group of minority children before conducting a race-comparative study and then attempt to generalize the findings to other non-white children.

It must be emphasized that it was not as if we were unaware of another level of inquiry via traditional scientific methods with random sampling, control groups, and so on. We specifically wanted to look at culturally diverse children in greater detail than is normally done by researchers. We strongly believe it is faulty methodology to use Anglo-American children as the point of reference for all children in the United States.

Another issue was that it was necessary to examine reliability and validity factors not only in "stepping away" from standardized procedures but also in the development of the procedures for potential assessment. On the basis of the information using primarily the Wechsler scales, we have developed a standard procedure for potential assessment on several commonly used standardized IQ tests—the Wechsler scales, Woodcock, the Kaufman, and the Stanford Binet.

Future Research and Practice of the Biocultural Assessment System

Research is still being conducted with Anglo-American children and children residing in countries around the world to see if there is a significant difference in their performance when potential and ecological assessments are conducted. In addition, there is still concern about the item equivalency measure because validating cultural equivalence is extremely difficult. After an abundance of assessments, we conclude that it is not statistically possible to quantify cultural equivalence. Powerful information can be obtained clinically, however. Thus, psychologists who consider themselves clinicians and not psychometricians will still find this measure very beneficial.

Exploration of the other intelligences, in particular the interpersonal and intrapersonal areas, is still needed.

In general researchers should be mindful that when examining intelligent behavior, the task should have at least two attributes—cognition embedded within the items and the content base modality (verbal, spatial, auditory, motor, and kinesthetic) these cognitions are tapping.

Are Psychologists Prepared to Go Beyond the Role of a Psychometrician?

To gain a healthy appreciation for this biocultural assessment system, we often asked our workshop participants and graduate students in training to differentiate between a psychometrician and a psychologist. Generally, the responses reflect a clear understanding that the psychologist is a clinician and a diagnostician, which means that he or she must go beyond IQ tests to assess intelligence, whereas the psychometrician relies purely on standardized tests of intelligence to determine a child's intellectual functioning. Interestingly, despite this awareness, individuals who call themselves psychologists continue to rely solely on IQ tests because "it is a sin to step away from standard procedures." The challenge facing psychologists today is whether they are prepared to expand themselves beyond the IQ guild to embrace a more comprehensive approach to intellectual assessment—an approach that would afford them the opportunity to be more accurate in their assessment, diagnosis, placement, and ultimate treatment of children throughout the system.

According to our four-tier biocultural assessment system, psychometricians perform 25% of the work of psychologists (see Chapter 6). In other words, as far as we are concerned, where a psychometrician's work ends is the beginning point for a psychologist. We believe that those who continue to rely on standardized tests of intelligence to determine intellectual functioning really have a genetic predisposition to intelligence because the basis on which these tests is built is that intelligence is fixed and immutable. It is time for us to come straight out and let the public know where we stand. Are you a psychologist or a psychometrician? If one is a psychometrician, then like Herrnstein and Murray (1994) and Jensen (1979), such an individual believes that IQ is fixed, that it is explained more through its biological genetic structures, and that for the most part it determines success in the real world. If one is a psychologist, then he or she should understand the importance of recognizing the role of experience and context in a child's intellectual development. Such an individual endorses a more environmental, cultural explanation for a child's intelligence and believes that IQ is labile and can change with the appropriate intervention.

A Note to the Users of the Biocultural Assessment System

As you embark on this comprehensive approach to assessing children's intelligence, you will come to gain more confidence in your assessment skills. If you came out of a traditional mode of assessment, as most of us did, initially you may feel awkward adopting this new approach. As you gain more practice, however, you will experience great fulfillment in your ability to assess children more accurately and to avoid misdiagnosis, misinterpretation, and misplacement of children. It is a journey that when taken totally transforms you in such a way that you will never resume assessment in the manner you did previously.

You should be proud of yourself for your willingness to grow and question your previous educational experiences. This indeed requires great courage and much risk. Just when you think you understand it all, you are faced with the challenge of a child who comes from another culture that you may never have come into contact with in the United States. This demands that you grow more. Through that growth, however, you will be rewarded with the joy of knowing that you have a skill that can apply to any child from any part of the world. You will come to know that assessment is not something you do to get a result: it is an action that, in itself, gives you a glimpse of the future of that child and, therefore, fulfills its own purpose at each moment. You have the ability to make a tremendous difference in the life of a child. Remember that one child who is misassessed is one child too many. Therefore, please choose your assessment tools wisely.

6

The Biocultural Assessment System

Preassessment Activities to Conducting the Biocultural Assessment System

Before assessing any child and before administering any battery of tests, a differential diagnosis of other possible causes for the child's problems should be performed (Armour-Thomas & Gopaul-McNicol, 1997a). Only after ruling out the following possible causes of the child's learning or emotional difficulties should a psychometric or nonpsychometric assessment be done:

Health assessment: It is important to review the child's school records to determine that all is well physically. In addition, ask the parent about the child's medical history. This is important to ensure that the child is not suffering from dietary deficiencies or any other ailments that can impede his or her functioning on the testing situation. The health examination must be done by a licensed physician or evidence of such in the form of a health certificate. This is important to rule out basic impediments to learning. This includes sensorium functioning: vision and hearing. Dental check-up should also be done to rule out the possibility of dental pain being a depressing factor. Blood work should be done to ensure the child is not anemic or suffering from dietary deficiencies. The issues of pain and anxiety related to menstruation should be explored. Finally, issues related to enuresis should be examined.

Linguistic assessment: It is necessary to rule out linguistic issues that may be the contributing factor to cognitive delays—that is, does this child speak another language other than English and, therefore, does not fully comprehend what is being said to him or her?

Prior experiences: It is necessary to examine any educational or psychosocial previous experiences, such as the child's learning style, that may inhibit or facilitate the expression of intellectual behavior. Also, it is important to investigate whether the child was formally educated. Many children who come from politically unrested countries may never have been formally educated. As such, they may be educationally deprived even though they may show similar profiles to those of mentally deficient children.

Family issues: It is advised to explore what familial factors, such as a recent divorce, may be affecting the child's performance in the clinical or school setting.

After ruling out any other causes for a child's learning and emotional difficulties, the four-tier biocultural assessment system then follows in the sequence outlined in Table 6.1. Table 6.2 summarizes for the examiner a step by step view of the stages in conducting this comprehensive assessment.

Biocultural Approach to Intellectual Assessment

The biocultural approach to intellectual assessment emphasizes that behavior is "intelligent" to the extent that the nature and quality of experiences to which one is socialized require the exercise of these capacities in a given context. This assessment system comprises three interrelated and dynamic dimensions: (a) a set of biologically pro-

TABLE 6.1 The Biocultural Assessment System

Psychometric Assessment

Psychometric Potential Assessment

This tier consists of the following four components:

Suspending Time
Contextualizing Vocabulary
Paper and Pencil
Test-Teach-Retest

This section reveals the child's potential and estimated intellectual functioning. If the child showed an improvement in his or her performance, the examiner should state so.

Ecological Assessment

This ecological taxonomy of intellectual assessment consists of the following four components:

Family/Community Support Assessment
Observation to determine performance in the school, home, and community (item and cultural equivalence)
Stage of Acculturation
Teacher Questionnaire

These components are used to assess the child in the following three settings:

School (classroom, gym, and playground)
Home
Community (church, playground, other recreational sites, or all three)

In this section, a child is observed in his or her ecology—home, community, and school. Therefore, the examiner discusses all tasks that the child was able to perform in these settings but that he or she was unable to do in the IQ testing situation, even under potential IQ assessment.

Other Intelligences

This tier consists of the following four components:

Musical Intelligence
Bodily Kinesthetic Intelligence
Interpersonal Intelligence
Intrapersonal Intelligence

grammed cognitive processes or capacities similar to those tapped by traditional standardized tests of intelligence (e.g., capacities for reasoning, auditory processing, and retrieving information from long-term memory); (b) experiences that mediate the use of cognitive processes under consideration. These experiences vary in a number of ways, including organization, form, content, and degree of famil-

TABLE 6.2 Sequence for Administering the Biocultural Assessment System

The Biocultural Assessment must be done in the following order:

1. Do a Differential Diagnosis by first looking at the following:
   Review school and clinic records: Secure the child's medical history.
   Teacher interview: Ask about the child's medical history, linguistic, other intelligences, and teacher questionnaire.
   Parent interview: Interview the parents at school or in the clinic. Ask about the child's medical history, other intelligences, and conduct the family and community support assessment to ascertain linguistic, educational experiences, and family issues.
2. Assess the child's psychometric intelligence in school. You must have two scores: one for the standardized questions and one for the potential questions.
3. Assess the child ecologically by observing the child in the home and community.
4. Conduct a parent interview for further ecological assessment of the child in the home and community.
5. Conduct a teacher interview for further ecological assessment of the child. Observe the child in the classroom and playground.

iarity and complexity; and (c) cultural niches within which these experiences are embedded and that function to enable or constrain the deployment of cognitive processes. Ecological contexts include the home, the community, and various settings within the school (e.g., classroom, playground, and cafeteria). The assumption of the assessment system is that intellectual behavior will vary within and between cultural groups insofar as there are differences in the experiences that different ecological contexts enable or impede the application of biologically constrained cognitive processes. Therefore, the cognitive capacities required for intelligent behavior in one context may or may not be the same as those in another context (Armour-Thomas & Gopaul-McNicol, 1997b; Gopaul-McNicol, 1992a, 1992b; Gordon & Armour-Thomas, 1991; Neisser et al., 1996). Thus, to assess intellectual functioning fully, a comprehensive assessment system is required that appraises how well an individual or group applies different cognitive processes for any given experience across multiple contexts.

We propose a more flexible and ecologically sensitive assessment system that allows for greater heterogeneity in the expression of intelligence. Our four-tier biocultural approach outlined in this book incorporates both quantitative and qualitative information with respect to cognitive functioning through various modes of assessment that include (a) psychometric, (b) psychometric potential, (c) ecological taxonomy, and (d) other intelligences (see Table 6.1).

Biocultural Assessment System

Psychometric Assessment

The important point to remember is that there is no single psychometric measure that taps the three interrelated and dynamic dimensions of intelligence—biological cognitive processes, culturally coded experiences, and cultural contexts. Therefore, any psychometric measure or an amalgamation of tests (interbattery testing, the process approach to assessment, or cross-battery testing) that emphasize a score-oriented approach should be used in conjunction with nonpychometric ecological measures because they help to further gain an understanding of the child's potential intellectual functioning and his or her ability to function in other settings besides the school.

It is critical to emphasize that this assessment system gives pure psychometric assessment (biological explanation) only 25% of the entire weight for determining an individual's intellectual functioning (Figure 6.1). Psychometric potential, ecological assessment, and other intelligences are each weighted 25% as well. Thus, the biocultural assessment system relies more heavily (75%) on one's experiences nested within one's contexts in determining one's intelligence.

Given the breakdown in Figure 6.1, one will expect that only 25% of the diagnostic power comes from standardized tests of intelligence. Chapter 8 demonstrates this via a case study. Prescription and intervention strategies are directly formulated from the information gained through potential and ecological assessment (see Chapter 9 for more detail on diagnostic and prescriptive utility).

Psychometric Potential Assessment

The advantage of assessing a child's potential during the testing process itself is that one is able to witness the improvement in a

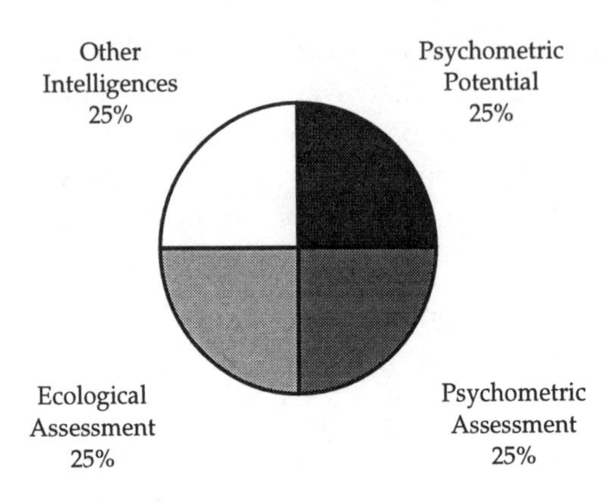

Figure 6.1. Four-Tier Biocultural Assessment System

child's performance immediately as opposed to waiting 3 years to see the gains when nonpotential psychometric assessment is used (Vygotsky, 1978). The psychometric potential assessment procedure consists of five measures that should be used in conjunction with the psychometric measure. They are used to provide supplementary information on cognitive functioning that goes beyond what is provided by the traditional standardized measure of intelligence. A description of each procedure is provided in the following sections.

Suspending Time

The assumption that to be smart is to be quick permeates the entire American society. Thus, many contemporary theorists (Carroll, 1993; Eysenck, 1982; Horn, 1991; Jensen, 1979; Woodcock, 1990) based their theories on individual differences in the speed of information processing and viewed speed as a major correlate of general intelligence. This assumption also underlies the majority of creative tests for gifted children. Several researchers differed with these researchers' positions and argued that there is still much doubt and uncertainty with respect to reaction time and psychometric intelligence (Carlson, 1985; Jones, 1985; Nettelbeck, 1985). Das (1985) noted that many American blacks have surpassed whites in the judicious use of their speed ability, especially in athletics and dancing, but on speeded tests they do not do as well as whites. The explanation must be in their familiarity or lack thereof of certain stimuli. Sternberg (1984) argues that although speed may be critical for some mental operations, "the issue ought not to be speed per se, but rather speed selection: Knowing when to perform at what rate and being able to function rapidly or slowly depending on the tasks or situational demands" (p. 7). Sternberg (1984) also argues that although speed of mental functioning has been associated with intelligence testing, it is well-known that snap judgments are not an important attribute of intelligence. Thurstone (1924) emphasized that a critical factor of intelligence is the ability to substitute rapid impulsive responses for rational, reflective ones. Noble (1969) found that children can be taught to increase their reaction time. Jensen and Whang (1994) agreed with Noble because they found that "the more the retrieval process has become automatic through practice, the faster it occurs" (p. 1). Therefore, the greater the speed, the greater the amount of practice. Baron (1981, 1982) also noted that with respect to problem solving, a reflective cognitive style is generally associated with intelligence. De Avila and Havassy (1974) noted that assessing culturally different children via timed tests confuses the measurement of ability with measurement of aspiration because little regard is given to children who are not culturally trained to work under timed conditions. Gopaul-McNicol (1993) found that most Caribbean children have difficulty completing tasks under time pressures because this represents the antithesis of what their culture dictates. On the contrary, slow and careful execution of their work is highly valued so that even if the child is aware that he or she is being timed, he or she may ignore the request by the examiner for a quick response and will rather execute the work methodically and cautiously. As such, scores tend to be lower for such students on timed tests, which comprise most of the nonverbal subtests. Of course, there are some professions, such as air traffic controller, in which one must consequentially make quick decisions as a part of one's daily life. Although speed may be essential in such situations, most of everyday life's events do not require decision making in a few seconds typically allotted for problem solving on IQ tests. The important issue here should not be one of total time spent but time distribution across various kinds of processing and planning events. The practical point to be made from this is that students should not be penalized for not completing a task in the allotted time.

Instead, they should be credited for successful completion of the task. Again, two scaled scores can be tabulated to compare how they function under timed conditions and how they function when tested to the limits.

Thus, this measure involves the suspension of time and the tabulating of two scores—one timed and one in which time is suspended.

Contextualization Versus Decontextualization

Although McGrew (1995) found that vocabulary is only moderately influenced by American culture, Hilliard's (1979) question, "What precisely is meant by vocabulary?" is a valid one which advocates for IQ tests have not yet answered. Words may have different meanings in different cultures. For instance, although the word *tostone* means a quarter or a half dollar to a Chicano, it means a squashed part of a banana that has been fried to a Puerto Rican. Given such a situation, it is recommended that the child be permitted to say the words in a sentence to be sure that the child's understanding of the word meaning is the same as that on the American IQ test. Armour-Thomas and Allen (1993) found that 32 ninth-grade students' vocabulary were elevated when unknown words were presented in a context-embedded situation. These findings were consistent with those of other studies that found individual differences in the acquisition of word meanings in contextually embedded situations (Sternberg, Powell, & Kaye, 1982; Van Daalen-Kapteijns & Elshout-Mohr, 1981). The important issue here is that most of vocabulary is contextually determined—that is, it is learned in everyday contexts rather than through direct instruction. Children accomplish this decontextualization by embedding unknown words in simple contexts (Sattler, 1988; Sternberg, 1985a).

In the authors' private practice, we found that children who did not know the word meanings in isolation were able to figure out the words when placed in a surrounding context. Of course, on traditional IQ tests children are asked word meanings in isolation. Although this may be acceptable for children who have had adequate educational opportunities in adequate social environments, for children who have had little formal schooling, word definition without the surrounding context may lead to invalid findings of their intelligence, in particular knowledge acquisition. Gardner (1983) recommended against using formal instruments administered in a decontextualized setting but instead recommended that assessment should be part of the natural learning environment and should not be set apart from the rest of the classroom activity.

With the biocultural assessment system, the examiner can contextualize all words by asking the child to say them in a sentence. For example, the examiner can say to the child, "Please say the word clock in a sentence." Potential credit is given only if the child says it in a sentence (not the examiner).

Paper and Pencil on the Arithmetic Subtests

During the past two decades, researchers have studied problem solving in mathematics from a cognitive information-processing perspective and found that a primary source of difficulty in problem solving lies in students' inability to monitor and regulate the cognitive processes that one engages in during problem solving (Artzt & Armour-Thomas, 1992). On most IQ tests, arithmetic taps skill, memory and attention, and speed. In the standard procedure, it is difficult to tell which is operating. Potential testing allows the examiner to rule out which factor is operating. For potential testing on the arithmetic subtest of the Wechsler scales, the examiner can say to the child who fails, "Please use this paper and pencil and try to solve the problem." This response will fall under a potential score.

Test-Teach-Retest Assessment Measure

Although Esquivel (1985) emphasized that "performance scales of standardized intelligence tests appear to have the greatest predictive validity for Limited English Proficient students, and may provide a more accurate estimate of their actual abilities" (p. 119), the nonverbal subtests, contrary to the claims that have been espoused, are not culture fair and are definitely not culture free. In fact, it is "the information (direct experience) components of these tests that carry their culture bound characteristics" (Cohen, 1969, p. 840). Nonverbal tests rely on one's ability to reason logically. In some respects, they embody more analytic mode of abstraction than the quantitative information components. This is because at times the task requires the individual to extrapolate and relate to relevant parts of the test items. Thus, the manner of cognitive organization is relevant for successful performance on nonverbal tests. The Block Design and Object Assembly subtests are highly influenced by the American culture, and individuals exposed to such items will find the experience less novel and thus their performance will be more automized. Hence, the tests will not be measuring the same skills across cultures and populations. Most children who are from rural areas in Third World countries have had little if any prior exposure to puzzles and blocks. Sternberg (1984) emphasized that "as useful as the tests may be for within group comparisons, between group comparisons may be deceptive and unfair for nonverbal subtests" (p. 10). A fair comparison between groups would require equitable degrees of familiarity or novelty in test items as well as comparable strategies. Sternberg (1984, 1985b, 1986) found that it was the ability to deal with novelty that was critical to measuring subjects' reasoning skills. Gopaul-McNicol (1993) found that in working with Caribbean children, other more complicated activities that also measure nonverbal abstract reasoning and visual integration—as do the Block Design and Object Assembly subtests, respectively—and are more relevant to the children's cultural experiences should be considered. The average child who comes from such countries is very handy and is able to help in constructing buildings, making furniture, creating a steel pan, maneuvering a motor boat, or cutting grass with a cutlass even though he or she has no formal education in these areas. These tasks are as or more complicated than putting blocks or puzzles together. Therefore, it would not be logical to label these children as delayed intellectually when they have honed other more complicated nonverbal skills. Evidently, their American counterparts are not labeled as deficient because they are unable to perform some of the previously mentioned activities that these children can so easily do. These skills, however, are not measured on the typical Anglo intelligence tests. Gardner (1983, 1993) noted that the performance gap between students from Western cultures versus those from non-Western cultures narrowed or even disappeared when familiar materials were used, when revised instructions were given, or when the same cognitive capacities were tapped in a form that made more sense within the non-Western context. Thus, nonverbal tests have not been freed from their culture-bound components. Clearly, the substantive information experiences are still culture bound. When testing the limits of culturally different children on the nonverbal subtests, it is quite common for students to answer the more difficult items correctly after they have passed their ceiling points or after time limits have been expended. It seems as if the children learn as they go along, and that lack of familiarity may have been why they did not do as well on the earlier items. Unfortunately, by the time they understand how to manipulate the blocks and put the puzzles together, it is time to stop those particular subtests because the children have already reached their ceiling point. Of course, in keeping with standardization procedures, one should not receive credit for items passed after the ceiling point has been attained.

Feuerstein (1979, 1980) produced evidence of the plasticity of the human organism that has made cognitive performance modifiable through mediated learning experiences. Through his Learning Potential Assessment Device (LPAD), Feuerstein found that a substantial reservoir of the abilities of Jewish children remained untapped when traditional assessment instruments were utilized to determine the intelligence of these children. This LPAD instrument, as the name implies, involved a radical shift from a static to a dynamic approach in which the test situation was transformed into a learning experience for the child. The focus was on learning rather than on its product and on the qualitative rather than on the quantitative dimensions of the individual's thought (Feuerstein, 1980, 1990). Many researchers (Beker & Feuerstein, 1990; Budoff, 1987a; Feuerstein et al., 1986b; Glutting & McDermott, 1990; Lidz, 1987, 1991; Missiuna & Samuels, 1988; Vygotsky, 1978) suggest that the best way to predict learning efficiency is to assess it in an actual learning and teaching situation. Thus, dynamic assessment links testing and intervention with the goal of enhancing a child's performance through a particular intervention. The objective of this approach is to identify obstacles that may be hindering the expression of a child's intellectual functioning and then to specify the conditions under which the child's intelligence may be enhanced. In summary, the child's modifiability is an important outcome of this dynamic approach to assessment.

Unlike in standardized testing conditions in which the examiner is neutral, the test-teach-retest approach proposed in the biocultural assessment system allows the examiner to be interactive, and his or her interactions are an integral part of the assessment process. The ultimate goal is to link the assessment findings directly to the development of individualized educational intervention programs.

The Test-Teach-Retest Assessment Measure is to be administered only if the examiner realizes that the child was not exposed to these types of items prior to the testing—that is, if the child never played with blocks, puzzles, and so on. Then, the examiner is to teach the child and then retest him or her. For instance, on the Block Design, Picture Arrangement, or Object Assembly subtests of the Wechsler scales, if a child fails the items on both trials, for potential psychometric assessment the examiner can teach and give the test again. Credit is given only under potential if the child gets it correct after the teaching period. The important point to remember is that the exact procedures are followed as in the standardized testing, except time is suspended, teaching is done, and potential scores are given after the child passes the teaching items.

In addition, please try to answer the following questions:

1. How much did the child benefit from the training intervention?
2. How much training is needed to raise the child's performance to a basic minimum level?
3. How well did the child retain the skills learned in the training period?
4. How much more training is needed to ensure that the child retains what he or she learned?
5. How well does the child generalize to other settings (home) what he or she has learned?
6. How easily is the child able to learn other difficult problems different from what he or she learned in training?

Ecological Taxonomy of Intellectual Assessment

Westernized thinking is indeed ethnocentric in its assumptions that Western education nurtures disembedded, context-free thinking. Although this may have been an attractive position years ago, there is now evidence that these cognitive processes are developed by various aspects of one's environmental experiences. Therefore, when a child is asked, "How are an apple and a banana alike" and he or she does not know the response but does know how a mango and a coconut are alike, then the crucial role of context in an individual's perception of the problem is ignored (Lave, Murtaugh, & de la Roche, 1984). Likewise, when an American Indian child has learned to develop speed by games such as a bow and arrow, the fact the he or she does not perform as well on the current speeded IQ tests in no way suggests that this child is limited in speed. Cultural taxonomies are risen out of one's cultural contexts. Therefore, one's cultural experiences and context are integral to the development of one's cognition. Culture dictates the amount of time a child will spend on a particular task. Therefore, the people of some former Soviet countries, Somalia, and Western Africa, who have to barter for food on a daily basis, tend to have a greater conceptual comprehension of volume because an error in bartering for a volume of rice could lead to suffering. Likewise, in many Third World countries, because there are no street signs a strong conceptualization of spatial orientation results. Accordingly, one can infer that the development of a specific set of skills can only occur within a specific cultural context in response to specific knowledge and experience. Thus, the implicit assumption that attributes are constant across place irrespective of the context in which one finds himself or herself is erroneous. The fact that an individual can perform one task very well may have little relevance for performing equally well another task that obviously entails the same cognitive ability albeit in different contextual settings. This is because it requires different types of values of attainment to respond to challenges in different environments. Several researchers (Lave, 1977; Murtaugh, 1985; Rogoff, 1978) found that competency in using arithmetic operations in carrying out everyday duties is not always predictive on standardized arithmetic tests, although they tap the same arithmetic operations. For example, Carraher, Carraher, and Schliemann (1985) found that "street children" in Brazil intuitively developed models of probability to serve as street brokers for lottery tickets. These same children, however, have difficulty applying these models to solve similar types of probability problems in the educational setting.

An examination of the literature on the consistency between IQ scores and real-life attainments calls into question the isomorphism between these two situations. For instance, the types of skills required for success on the Picture Arrangement and Similarities subtests of the Wechsler scales are similar to the deductive reasoning necessary for grocery shopping. Grocery shoppers tend to match prices, comparing how similar or dissimilar items are, as well as plan whether the volume of their purchase can fit in their refrigerator. As such, the goal is to allow for a few days of supply rather than a week. The previous examples confirm that there are many instances in which deficits in cognitive functioning disappear when the problem is couched in familiar terms or using familiar stimuli (Super, 1980). Therefore, cognition is indeed context sensitive and there exist multiple cognitive potentials instead of one cognitive potential or one central processor.

Because we have veered too far in the direction of formal testing, and in the light of these desiderata for new approaches to assessment, several researchers propose a more naturalistic, context-sensitive, and valid ecological mode of assessment (Ceci, 1990; Gardner, 1993; Vygotsky, 1978). This is not merely a call to regress to a subjective form of evaluation. There is no reason to feel less confident about such a thorough approach because reliability can be achieved in these ecological approaches as well. In fact, these nonpsychometric measures that are based on multiple assessment instruments in multiple contexts have more ecological validity than psychometric measures that were based on a child's functioning in a controlled testing situation. Another retort to this alleged objectivity of standardized formal tests is the fact that all tests are skewed toward a certain type of cognitive style. Thus, standardized tests are quite hostile and unfriendly to individuals who do not possess a blend of certain logical and linguistic intelligences and who are uncomfortable in decontextualized settings under impersonal and timed conditions. Correlatively, such tests are biased in favor of individuals who possess these strengths based on their prior cultural experiences.

Therefore, this "assessment view" seeks to connect school activities with after-school activities with emphasis on the individual's strengths (Figure 6.2). In other words, this approach calls for a broader menu of assessment options and an abandonment of the sophomoric mentality that relies on some type of rigid superficial conformity. A broader trained cadre of workers would make greater use of the many subsets of human talents by embracing this assessment approach. The reader should note that the authors are not

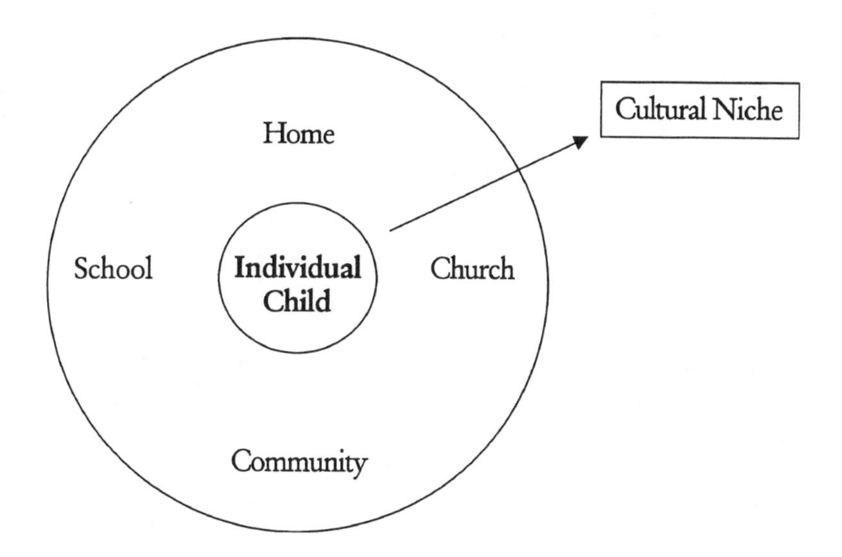

Figure 6.2. Biocultural Assessment System

asserting that psychometric tests are not relevant in the intellectual assessment enterprise. What is being asserted here is that assessment of intellectual functioning should be mindful of the multifaceted influences of culture on behavior.

Ecological Taxonomy of Intellectual Assessment

This ecological measure attempts to measure skills and behaviors that are relevant to the context in which a child lives (real-world types of intelligences, not just academic type of intelligence). Therefore, the child is assessed in several settings—the school, home, and the community. Observing children's interaction with their family and friends in their most natural settings brings to the assessment robust knowledge of the family dynamics and cultural experiences of the child. The examiner should look for

- The way they communicate
- The way they socialize
- The activities they engage in

- The friendships they have
- The roles they play
- The respect or lack thereof they are given by family and friends

In addition, the examiner should assess the child's intelligence by bringing some real-life experiences to the psychometric measure. For instance, if a child is unable to attain success on the mazes, take the child to a real-life maze situation and see if the child can maneuver his or her way out of the maze. Even intelligence experts such as Wechsler (1958) and Binet and Simon (1905) defined intelligence as one's ability to adapt to the real-world environment. Williams (1971) emphasized that the very fact that a child can learn certain familiar relationships in his or her own culture shows that he or she can master similar concepts in the school curriculum, as long as the curriculum is related to his or her background experiences.

Another example is the comprehension subtest; if the child does not know what he or she is supposed to do, then take him or her to a real situation and see if he or she is able to perform the task. For example, take the child to a store and see what he or she will do if he or she finds a wallet that you deliberately placed.

The second component within the ecological taxonomy is the Family/Community Support Assessment. This is a questionnaire designed to determine what support systems the child has at home or in the community, what has been the child's previous educational experiences, what language is spoken at home, and how the family can boost a child's intellectual functioning. Family assessment as part of intellectual assessment brings to the assessment robust knowledge of the family dynamics of the child. Therefore, parenting, child-rearing practices, disciplinary measures, punishments and reinforcers, the language spoken at home, religious values, the child's relationship to the society at large, as well as society's impact on the child are indeed rich sources of knowledge for school psychologists.

There is an enormous amount of literature that shows the relationship between individual differences in intellectual functioning and individual differences in familial support (Nichols, 1981; Zajonc, 1976). The basic proposition is that there is a direct relationship between the intellectual environment of the family and the intellectual development of the child being socialized in that environment.

Parenting has been shown to exert a powerful influence on intellectual development, primarily through the inculcation of specific modes of learning strategies and motivational routes that influence problem-solving skills (Vaughn, Block, & Block, 1988). Therefore, child-rearing practices such as encouraging proper study habits are congruent with later intellectual development. There is substantial literature that documents social class differences in parenting that are linked to differences in intellectual functioning. Schaefer (1987) suggested that intellectual competence is mediated by an assortment of parental attitudes and values, such as the worth of schooling, and so on. Schaefer's environmental model posits that human intelligence is not fixed but can be changed by altering parental attitudes and values. Furthermore, various aspects of home environment and parental values are related to IQ and educational attainment (Siegel, 1984). Others have shown that the home environment is a better predictor of educational outcomes than cognitive variables, IQ, and socioeconomic status (SES), even when SES was controlled in a regression analysis. These studies clearly show that aspects of the home, such as its organization, are important independent determinants of intellectual development.

Family assessment has become a point of discussion in the school setting. This is because to really aid in promoting psychological well-being in children, individual assessment will have to be done in conjunction with family assessment, particularly with respect to at-risk families. Thus, the kind of support systems students have at home should be rated and ranked, as is done with other intellectual assessment measures. Thus, a student who is found to be of low average intelligence on an IQ test but has great family support has the potential to be an average to above average student. This is indeed a very critical issue for determining the potential, motivation, and achievement of a student. Likewise, very intelligent students who are not achieving commensurate to their ability may need to be given more familial support. Moreover, the relationship between parents' schooling and the intellectual functioning of their children suggests that public policymakers can exert a positive influence on the intellectual performance of children. By encouraging parents, especially those from culturally different backgrounds, to advance themselves educationally and to become more involved in their children's education an improvement in the intellectual performance of their children can result (Valencia, Henderson, & Rankin, 1981).

Interpersonal competence must also be examined with respect to the child's culture because ethnic groups vary in the value attached to certain kinds of interpersonal skills (Taylor, 1988). Certain behaviors that are reinforced in one group may not be in another. For instance, in the white American culture, children who are assertive are reinforced, whereas among African Americans and immigrant groups such behaviors are considered disrespectful and hence are not reinforced. Likewise, among many culturally different families, parents place different emphasis on independence at an early age. Adolescent autonomy is often not reinforced in such societies. Thus, the evaluator or clinician must be knowledgeable about the ethnic values toward this issue to more accurately determine whether the family respected its own cultural norms with respect to autonomy in children. Likewise, for children who are not assertive or are shy in the classroom, it would be wise to conduct a family assessment to determine if such behaviors are in fact reinforced at home. To encourage a child to completely abandon his or her familial cultural values is ill-advised given the disorientation this can cause to both the child and his or her family. It would be beneficial to assist the family in understanding what cultural adjustments the child is making to adjust to the school or community. This awareness should aid in the family's acceptance of some of the "strange" behaviors evidenced in their children.

Family/Community Support Assessment

Parent's name: ______________________ Date: __________

Child's name: ______________________________

Regionality: USA—urban/inner city/suburban/rural

If country of origin is not in the USA: ______________________

Where is the native country? ______________________

How long is the child in the USA? ______________________

Are the parents/significant other residing in the USA? __________

Who else resides in the home? ______________________

What has been your child's previous educational experiences? ________

____

____

- Was your child ever retained? ______________________
- How often per week is your child absent? ______________
- What has been your child academic performance in

  Math—poor/fair/good/very good

  Reading—poor/fair/good/very good

- Did your child participate in any supplemental instructional programs? If yes, what programs were they? ____________

In the event there is one parent in the home, but there is a significant other residing at home, fill in the other relative instead of mother/father. If there is only one adult in the home, then leave the other parent section questions blank. ______________________________

How many years has the mother been residing in the United States?

____

How many years has the father been residing in the United States?

____

What is the mother's place of birth? ______________________

What is the father's place of birth? ______________________

____

Home Linguistic Assessment

1. Does mother speak English? Rank how well.
   Not fluently Somewhat fluently Very fluently
2. Does mother speak another language? What language?
   Not fluently Somewhat fluently Very fluently
3. Does mother read English? Rank how well.
   Not fluently Somewhat fluently Very fluently
4. Does mother read another language? What language?
   Not fluently Somewhat fluently Very fluently
5. Does mother write English? Rank how well.
   Not fluently Somewhat fluently Very fluently
6. Does mother write another language? What language?
   Not fluently Somewhat fluently Very fluently
7. Does father speak English? Rank how well.
   Not fluently Somewhat fluently Very fluently
8. Does father speak another language? What language?
   Not fluently Somewhat fluently Very fluently
9. Does father read English? Rank how well.
   Not fluently Somewhat fluently Very fluently
10. Does father read another language? What language?
    Not fluently Somewhat fluently Very fluently
11. Does father write English? Rank how well.
    Not fluently Somewhat fluently Very fluently
12. Does father write another language? What language?
    Not fluently Somewhat fluently Very fluently

In what language did the mother receive most of her education?

____

In what language did the father receive most of his education?

____

What language does your child most often speak to his or her mother?

____

What language does your child most often speak to his or her father?

____

What language does your child most often speak to his or her siblings?

____

What language does your child most often speak to his or her friends?

____

In what language are radio or television shows most often received at home?

____

How many hours per week does your child read? ______________

In what language does your child most often read? ______________

How many hours per week does your child see you reading? ________

In what language does your child most often see you reading? ________

Linguistic proficiency: ______________________

Linguistic dominance: ______________________

Family Educational Background

13. Mother level of education attained in native country or in the USA

    Did not finish HS Finish HS Finish college

14. Father level of education attained in native country or in the USA

    Did not finish HS Finish HS Finish college

15. Is anyone at home able to assist this child in his or her homework?

    Never Sometimes Always

16. Is the child able to study at home?

 Never Sometimes Always

 Why not? ______________________________

17. Is anyone or organization in the community able to assist the child in his or her homework?

 Never Sometimes Always

Social/Community

18. Are you involved in any church/community organization?

 Never Sometimes Always

19. Is your child involved in any church/community organization?

 Never Sometimes Always

20. Are you involved in any interest group or club?

 Never Sometimes Always

21. Is your child involved in any interest group or club?

 Never Sometimes Always

22. Does your child go to any nonschool events with friends without adult supervision?

 Never Sometimes Always

23. Does your child go to any nonschool events with friends with adult supervision?

 Never Sometimes Always

24. Is your child involved in any sports in the community?

 Never Sometimes Always

25. Is your child involved in any form of employment?

 Never Sometimes Always

Examine the circled responses and comment on the level of support based on these answers, the clinical interview, behavioral/cultural observation, etc.

More "never" responses—low

More "sometimes" Responses—moderately low/adequate

More "always" Responses—adequate/moderately high

Furthermore, to help the examiner to conduct the item equivalency assessment, it is necessary to ascertain from the parent or significant other how the child learned the common IQ constructs in his or her culture. Therefore, the following should be stated or asked of the parent:

Please assist me in understanding how your child learns or what type of experiences he or she had in his or her native country or culture by answering the following questions:

How are the constructs listed below taught or nurtured in your culture—that is, what games do children play, what songs do they sing, or what social experiences do they have to help develop these skills?

(These may have to be simplified to fit the parents' educational level)

Verbal abstract reasoning—similarities: How do children learn the differences and similarities in objects or items?

Social awareness—comprehension: How are children taught to be aware of social cultural issues (directly or indirectly)?

Word meaning—vocabulary: How do children learn new words? How is vocabulary developed?

Arithmetic: Check with the parents to find out what level of arithmetic skills the child attained commensurate to his or her age. (The examiner can allow the child to do paper and pencil tasks to assess his or her potential.)

General information: How is general information learned—read papers, teach civics?

Auditory short-term memory—digit span: How is memory developed—songs, rhymes, and history?

Visual discrimination—picture completion: How are children taught to pay attention to detail?

Visual integration—object assembly: How are children taught to build things up—puzzles, etc.?

Nonverbal comprehension/planning—picture arrangement: How are children taught to plan, organize, think sequentially, follow directions, etc.?

Nonverbal abstract reasoning—block design: How are children taught to break something down and build it back up?

Visual short-term memory—coding: How are children taught to memorize things they see in a short time?

Item Equivalencies Assessment Measure

The third component within the ecological taxonomy is the cultural and item equivalence. In some ways, standardized tests of intelligence are consistent with a biocultural approach. They all purport to measure many cognitive abilities indicative of intelligence and that many scholars agree are, in part, biologically determined. Test items do reflect learning experiences that are similar to the learning experiences common in some homes and in some school contexts. The caveat, of course, is that the experiences sampled on the tests are not common across populations. As such, it is indeed inappropriate to suggest that there is a resolution with respect to interracial or cultural equivalence on intelligence tests, and it is equally misleading to conclude that one is brighter by virtue of his or her score on these IQ tests. Therefore, existing tests should incorporate multiple methods that examine both biological and ecological factors.

Cummins (1984) points out that when referring to concept formation, evaluators must keep in mind that it is difficult for examinees to know similarities or differences in objects if they have had little or no experiences with the objects themselves. An alternative would be to determine if these children can perform comparable skills typical of their native lands and describe in a more qualitative type of manner their strengths and weaknesses with respect to the skills they can perform in their native countries, showing the similarities between these skills and those found on the traditional-type tests. For instance, Question 4 on the Similarities subtest of the Wechsler Intelligence Scale for Children—III reads, "In what way are a piano and a guitar alike?" Many children from Third World countries may not have ever seen or heard a piano. Perhaps the cuatro, another string instrument, could be substituted. The same solution may be applied to Question 6—"In what way are an apple and a banana alike?" Apples are not grown in tropical climates. Perhaps mango could be substituted. The important thing here is that the child knows the concept of fruits of different kinds. This idea of matching items to a child's culture has been emphasized by Sternberg (1986) and Helms's (1992) cultural equivalence perspective.

Many test developers have attempted to reduce or ignore cultural influences on cognitive ability tests by constructing "culture fair tests." McGrew (1994) spoke of the comprehensive nature of the Woodcock newly revised test of cognitive ability. These tests, however, are simply attempts to control the influence of different cultures instead of measuring them (Helms, 1992). Even the black IQ test that was developed by Williams (1975) was biased in favor of a specific social class and regional group rather than blacks as a cultural group. The insistence that white American culture is superior and universal and should be adopted by every racial group results in the devaluation of the unique and special cultural values of different groups. Few investigators have studied what blacks and other minorities have contributed to this society apart from being used as points of comparisons on IQ scores. Lonner (1981) discussed the following four types of equivalence:

1. Functional equivalence: the extent to which the test scores mean the same thing among different cultural groups and measure psychological characteristics that occur equally frequently within these groups
2. Conceptual equivalence: whether groups are equally familiar or unfamiliar with the content of the test items and therefore attribute the same meaning to them
3. Linguistic equivalence: whether the test developer has equalized the language used in the test so that it signifies the same thing to different cultural groups
4. Psychometric equivalence: the extent to which tests tap the same things at similar levels across different cultural groups

Butcher (1982) listed potential nonequivalent sources in cross-cultural research and emphasized that failure to consider these issues can result in committing the cultural equivalence fallacy. Helms (1992) pointed out the culture bias with respect to psychometric equivalence. Helms (1989) defined Eurocentricism as "a perceptual set in which European and/or European American values, customs, traditions and characteristics are used as exclusive standards against which people and events in the world are evaluated and perceived" (p. 643). Thus, permeating every question of these IQ tests is whether the answers are right or wrong, with the correctness being determined by the normative white response. Thus, the more intelligent individual is the one who can apply Eurocentric rules most effectively and expediently. Therefore, those who see alternative answers because they do not have access to the Eurocentric worldview are penalized and are deemed less intelligent. Matthews (1989) reported the incident when two white prelaw students convinced the Law School Admissions Services that a Law School Admissions Test question had two correct answers. Although these two men were socialized in a white culture, they saw an alternative and more creative response than did the test's constructors. Another example could be found in Gopaul-McNicol (1993). The author noted that many "lesser developed" cultures value functional responses rather than taxonomic responses because it is in keeping with the normal everyday life for many groups. Therefore, when a child gives a one-point response to "How are a car and a boat alike?" by saying that "you drive them," it is in keeping with the functional use of these items rather than the taxonomic response of them being vehicles or means of transportation. As Piaget (1952) pointed out, however, the taxonomic responses are considered more abstract and therefore show a more advanced form of cognitive ability.

Until test developers investigate the item equivalence on these tests, the lack of cultural equivalence on the IQ tests cannot be ruled out as an explanation for the between-group racial difference of IQ scores. Until test developers assess whether black or African culturally laden cognitive strategies turn out to be more effective strategies for equal or better outcomes for predicting performance, it should not be assumed that blacks are less intelligent. A recommendation would be to allow expert representatives of various ethnic, cultural, gender, linguistic, and social class groups to assist in the construction of the test items. The caveat, of course, is whether these consultants have the ability to step away from the Eurocentricism so endemic in their professional environment. Please note that we are not asserting that psychometric tests are not relevant in the intellectual assessment enterprise. What is being asserted here is that other cultural groups centered cognitive abilities require test developers to integrate ecological contextual factors into their thinking process. In other words, these other group information processing strategies might be some sort of implicit unmeasured aspect of ability tests in addition to their predictive criteria. In many cultures throughout the world, teachers and parents do not always reward the obvious answer but rather reward the more creative, expansive responses. Thus, when an individual who is socialized to develop and nurture innovative, expansive, interactive, and spontaneous thinking is placed in a testing situation, it is difficult for him or her to reconcile the contrasting Eurocentric world perspectives that underlie test construction. In many British colonized societies, for instance, a student's achievement is not assessed by multiple-choice exam questions but by the ability to respond to essays, resulting in a more creative, integrative learning and response style. For these students, multiple choice is viewed as simplistic and not a mature way of assessing one's comprehensive knowledge (Gopaul-McNicol, 1993). Students who have to adjust to a more detailed way of studying to respond to the right and wrong multiple-choice format usually find this frustrating.

In general, different responses are elicited on the IQ tests depending on the environment in which the child was reared and depending on the culturally loaded biased items. An example of a biased item is from the comprehension subtests of the WISC-III: "What is the thing to do if a boy/girl much smaller than yourself starts to fight with you?" In discussing this question, Helms (1992) stated that researchers (Gordon & Rudert, 1979) contend that black children are taught that the appropriate response is to "hit him/her back." After testing black children, we often try to ascertain their cultural experiences by interviewing their parents. In response to why their child may have responded in an "aggressive manner" to that question on the comprehension subtest, black parents often say, "My child cannot come home and tell me a white kid beat him or her up." It seems that when the black child in the testing situation hears "boy/girl," the child thinks white boy or white girl, so they know they are expected to defend themselves as they are taught by their families. It is not so unusual for black parents to teach their children to self-defend given the racism in this society. The fact that the question recognizes the need to say "boy" if it is a male child being tested as opposed to "girl" if it is a female child being assessed tells us that the response would have been different had the examiner asked a girl what she is supposed to do if a boy hits her. Mothers often tell their daughters to never allow a boy or man to strike them. Thus, a girl would have probably given a more aggressive response if the question states that it is a boy doing the hitting as opposed to a girl. It is expected that black children's responses might be different if they think that it is a black instead of a white child doing the hitting.

This item equivalency assessment measure attempts to equate a child's cultural experience in every item of every IQ test by matching the questions on the IQ test to the child's culture. As such, the child's broad-base information repertoire is recognized. A caveat is that it is not statistically possible to quantify cultural equivalence. Powerful information can be obtained clinically, however. Thus, psychologists who consider themselves more than just a psychometrician will still find this measure very helpful because they can clinically create a cluster of items that form the construct of intelligence for a particular cultural group. This cultural equivalence approach certainly also falls under the rubric of potential assessment.

Item Equivalencies Assessment Measure

Because the Wechsler scales are the most widely used of all intelligence tests, a cursory review of the Wechsler items will be examined to demonstrate the ecological nonpsychometric measures.

If the examinee did not obtain full score points on any subtest, the item equivalency measure should be administered.

Creating a cluster of items that form the construct of intelligence that are equivalent items for the culturally or linguistically different child is one way of tapping a child's intellectual potential. Tabulating two scores—one following standardization procedures and one measuring the child's potential —should result in a more accurate assessment of the child's intelligence.

Nonverbal equivalencies: The parent interview should assist you in matching the child's cultural experiences to the test constructs. Thus, assess if the child has the concept and how it is manifested in his or her culture.

In addition to suspending time for all the timed subtests, the following items can be matched to the child's culture.

Tests Administered	*Subtest*	*Item Equivalency*
WISC-III	Gen inf.	Match to culture
Verbal subtests	Similarities	Match to culture
	Vocabulary	Match to culture or contextualize all or both
	Arithmetic	Match to culture (paper and pencil)
	Comprehension	Match to culture; place the question in different context
Nonverbal subtests	Block Design	Match to culture or test or teach; retest or see what other skills measure this concept
	Object Assembly	Match to culture; what skills measure this or teach
	Picture Completion	Match to culture; what skills measure this; if the child points, score as correct; you are not assessing vocabulary
	Picture Arrangement	Match to culture; what skills measure this or teach
	Coding	Match to culture; what skills measure this (e.g., memory and card game); people have different symbols systems in their culture

Review each item of each of the verbal subtests and match them to the child's culture. For example, on the Wechsler verbal subtests:

General Information

Question 9: "What are the four seasons of the year?" A child who recently arrived from a tropical climate, where there are only two seasons—rain and sun—may be at a disadvantage. If the child knows the two seasons in his or her native country but is unfamiliar with the four seasons in the United States, two general information subtest scores should be tabulated: one following standardization procedure and one showing the child's potential.

Similarities Subtest

Cummins (1984) points out that when referring to concept formation evaluators must keep in mind that it is difficult for examinees to know similarities or differences in objects if they have had little or no experiences with the objects themselves.

Question 4: "In what way are a piano and a guitar alike?" Many children from Third World countries may not have ever seen or heard a piano. Perhaps the cuatro, another string instrument, could be substituted.

Question 6: "In what way are an apple and a banana alike?" Apples are not grown in tropical climates. Perhaps mango could be substituted. The important thing here is that the child knows the concept of fruits of different kinds.

Question 9: "In what way are a telephone and a radio alike?" Children who are not exposed to telephones may never have thought about this, and in many Third World countries, only the affluent have telephones. Therefore, some children may not have conceptualized in an abstract sense a telephone as it relates to a radio.

Comprehension Subtest

Question 8: "What is the thing to do when a boy or a girl much smaller than yourself starts to fight with you?" Most children from different cultures may answer tell his or her mother or tell the teacher. From a ecological standpoint, this is quite expected and acceptable. This response, however, results in only one point. For such children to get two points it may be necessary to ask, "Would you fight?" Usually, the child will say, "No, it is not good to fight." The important thing to keep in mind is that in such cultures not fighting and telling an adult are the most important things taught to children. In this culture, a child is not necessarily expected to tell an adult.

Question 13: "Why is it good to hold elections by secret ballot?" Children originally from nondemocratic societies, such as Haiti, Cuba, and so on, may have difficulty with this question because the idea of a secret ballot is foreign to them. The very concept is difficult for such children to understand because in their societies voting is not a private matter and in some cases there is no such thing as an election.

Question 15: "In what ways are paperback books better than hardcover books?" In some cultures, such as the United Kingdom and the English Caribbean, the term softcover rather than paperback is used, so it is best to use an equivalent term with such children.

Change the contexts of the question. For example, for Question 4, "What are you supposed to do if you find one's wallet or purse in a store?" the examiner can change the context—for example, change "in a store" to "in a police station" or "in the classroom."

It is important to remember that this subtest is not attempting to assess what a child would do but rather what he or she is expected to do in a social situation. In other words, in what contexts would the child choose to do what he or she knows he or she is supposed to do.

Digit Span

Examine how else auditory short-term memory is assessed in the child's culture (parent interview). In addition, remember people have different symbol systems in their culture. Therefore, using the more familiar letters than numbers may facilitate recall.

Two additional questionnaires make up the fourth component of the ecological taxonomy. These two questionnaires were designed to further assist the examiner in understanding the child. The stage of acculturation of a child provides information about the child's emotional functioning that may be impeding intellectual functioning and classroom or home adjustment. The teacher questionnaire allows us to view the child through the eyes of the teacher and gives us important information about the child's abilities and overall functioning in the classroom. All this information is critical in grasping the total picture of the child vis-à-vis his or her entire community.

Stage of Acculturation

Assess what cultural adjustment difficulties the individual may be experiencing. This is done because in working with children, in particular culturally different children, to some extent most people undergo some change (minimal as it may be) at unpredictable periods of time:

1. Physical changes: The individual must cope with living in a new place in which pollution and other environmental hazards can be a new experience.
2. Cultural changes: Linguistic and social institutions are different and thus the individual has to adjust to these differences. The individual has to function within new social networks both within his or her own group and outside his or her group.

3. Psychological changes: The individual may experience an alteration in his or her mental status due to culture shock as he or she adapts to the new milieu. This is a period of psychological transition from back-home values to host-home values. Individuals begin to understand the host culture and feel more in touch with themselves.
4. Acculturated: The individual has adjusted to the new culture but values his or her cultural morays as well. Thus, he or she is bicultural.

Teacher Questionnaire

- What has been the child's previous educational experiences?
- Was the child ever retained?
- How often per week is the child absent?
- What has been the child's academic performance in
  Math: poor/fair/good/very good
  Reading: poor/fair/good/very good
- Did the child participate in any supplemental instructional programs? If yes, what programs were they?
- What is the child's motivational or attention levels in class?
- How persistent is this child?
- How does the child relate to his or her peers?
- How does the child behave in class? In other words, is the child reflective or impulsive?
- Is the child responsible? How so?
- Is the child disciplined? How so?
- Does the child prefer to study alone or in a group?
- Does the child prefer dim or bright lights?

Also ask the music/dance/sports/drama teacher the questions from the Other Intelligences questionnaire.

Other Intelligences Inventory

The Other Intelligences Inventory is the fourth tier of the biocultural assessment system. This tier supports Gardner's (1983, 1993) position that most children can excel in one or more intelligences.

Gardner's theories of seven multiple intelligences shares the view that the human mind is a computational device that has separate and qualitatively different analytic ways of processing various kinds of information. He provides evidence for what he terms "factors of mind." He maintains that there are many types of intelligences rather than one, as IQ tests claim. He takes the position that all children can excel in one or more intelligences. Gardner regards his theory as an egalitarian theory, and what is most important to him is not whether one child outperforms another on some skills but that all children's skills are identified. Therefore, one child may have a propensity for interpersonal skills, whereas another may have a propensity for numerical reasoning. These propensities are not borne because a parent exposed their child to situations and activities involving these intelligences but rather because children have "jagged cognitive profiles," somewhat akin to the learning disabled child who does well in one area and poorly in another. Of all the theoretical positions, Gardner's is the one most outside the family of traditional intelligence researchers. He developed his thinking partly because he objected to what he viewed as the domination of thinking about intelligence by a few theorists. Many have criticized Gardner's thinking, claiming that he does not examine the possibility that various cognitive abilities exist within each type of intelligence. His theory has been criticized by psychologists on the basis that it cannot be subjected to adequate testing. In certain educational quarters, however, massive gains in children's motivational level and sense of self have been used to validate Gardner's theory.

James Comer, a professor at Yale University and the director of the leading school reform program in the country, the School Development Program, proposes a theory of development that includes multiple pathways (Haynes, 1995). Although Comer did not address IQ as a construct, Comer suggests that the effects of schooling are very significant and the environment is primarily responsible for successful development along six developmental pathways: physical, moral, linguistic, social and emotional, psychological, and cognitive and academic. Comer also emphasized that there is a physiological component by way of energy and drive that influences one's development along each pathway. Comer was most emphatic in stating that supportive environments, in which adults manifest much caring and nurturing, are most influential in children reaching their potential.

The Other Intelligences Inventory (Gopaul-McNicol & Armour-Thomas, 1997b) attempts to capture four of Gardner's intelligences. Therefore, asking the child, the parents, and the teacher questions to ascertain the child's musical intelligence, bodily kinesthetic intelligence, and inter- and intrapersonal intelligence—areas that are not represented on any of the commonly used standardized IQ tests (see Chapter 8)—is critical in understanding the breadth and depth of a child's comprehensive intellectual abilities.

Interviewing several different persons (child, teacher, and parent) is necessary to add reliability to the child's description of his or her other intelligences.

Other Intelligences Inventory

Does your child have a strength in any nonacademic type area? ________

____

How do you define this as a strength? Give examples of such.

____

____

____

Musical Intelligence

Interview With the Child

Do you (make music) play a musical instrument? Yes/No

If yes, what instrument? ________________

How long have you been playing this instrument ________________

Do you sing? Yes/No

If yes, what kinds of songs? ________________

How long have you been singing? ________________

What level of proficiency have you attained?

Beginner Intermediate Advanced

Please check off if you can do the following:

Listen to a piece of music then create a song

Could I obtain a copy of your certificate/diploma/teacher feedback in music/singing?

Do you have samples/audiotapes of musical/singing performances? Y/N

Do you have samples/audiotapes of compositions? Y/N

Do you have samples of written/performed/composed songs? Y/N

Do you have lyrics of raps, songs, or rhymes that you wrote? Y/N

Did you compile discographies? Y/N

Interview With the Parent and Music Teacher

Does your child (make music) play a musical instrument? Yes/No

If yes, what instrument? ______________________

How long has your child been playing this instrument? ___________

Does your child sing? Yes/No

If yes, what kinds of songs? ______________________

How long has your child been singing? ______________________

What level of proficiency has your child attained?

Beginner Intermediate Advanced

Could you furnish me with a copy of your child/your student certificate/diploma/teacher feedback in music? ______________

Do you have samples/audiotapes of musical performances of your child? Y/N

Do you have samples/audiotapes of compositions of your child? Y/N

Do you have samples of written/performed/composed songs of your child? Y/N

Do you have lyrics of raps, songs, or rhymes that your child wrote? Y/N

Did you compile discographies of your child? Y/N

Bodily Kinesthetic Intelligence

Interview With the Child

Do you play any sports? Yes/No

If yes, what sport(s)? ______________________________

How long have you been playing this game? ______________

What level of proficiency have you attained? ______________

Beginner Intermediate Advanced

Do you dance/act/paint/draw? Yes/No

If yes, what type of dance? ______________________________

How long have you been dancing/acting/painting/drawing? _______

What level of proficiency have you attained?

Beginner Intermediate Advanced

Could I obtain a copy of your certificate/diploma/teacher feedback in sports/dance/drama? _________

Do you have videotapes of projects/demonstrations? Y/N

Do you have samples of projects actually made? Y/N

Do you have photos of hands-on projects? Y/N

Interview With the Parent and Sports/Dance/Drama/Art Teacher

Does your child play any sports? Yes/No

If yes, what sport(s)? ______________________________

How long has your child been playing this sport? ______________

What level of proficiency has your child attained?

Beginner Intermediate Advanced

Does your child dance/act/paint/draw? Yes/No

If yes, what type of dance? ______________________________

How long has your child been dancing/acting/painting/drawing?

____

What level of proficiency has your child attained?

Beginner Intermediate Advanced

Could you furnish me with a copy of your child/your student certificate/diploma/teacher feedback in sports/dance/drama/art?

____

Do you have videotapes of projects/demonstrations of your child? Y/N

Do you have samples of projects actually made by your child? Y/N

Do you have photos of hands-on projects of your child? Y/N

Personal Intelligences

Intrapersonal

A self-concept scale should assist in assessing self-esteem (intrapersonal) issues. In addition, ask the following:

- Do you have self-assessment activities/checklist?
- What outside hobbies or activities are you involved?
- What are your strengths and weaknesses?

Break down into intellectual, social, and affective.

Interpersonal

Social skills scales such as the Social Skills Rating Scale should assist in assessing interpersonal intelligence. In addition, examine the following:

- Peer group reports
- Videos, photos, or write-ups of cooperative learning projects
- Certificates or other documentation of community service projects
- Written teacher reports
- Written parent reports

Break down further into intellectual, social, and affective.

Furthermore, the following questions can be asked:

Child Interview

Are you involved in any church/community organization?

Never Sometimes Always

Are you involved in any interest group or club?

Never Sometimes Always

Do you go to any nonschool events with friends without adult supervision?

Never Sometimes Always

Do you go to any nonschool events with friends with adult supervision?

Never Sometimes Always

Are you involved in any form of employment?

Never Sometimes Always

Do you accompany your parent or any relatives to the store or any agency and serve as a translator for him or her?

Never Sometimes Always

Are you responsible for caring or for supervising your younger siblings while your parents/relatives are not at home?

Never Sometimes Always

Could you cook/iron/go grocery shopping? Please circle the ones you do well.

Please list any other domestic/community chores for which you are responsible.

Parental Interview

Is your child involved in any church/community organization?

Never Sometimes Always

Is your child involved in any interest group or club?

Never Sometimes Always

Does your child go to any nonschool events with friends without adult supervision?

Never Sometimes Always

Does your child go to any nonschool events with friends with adult supervision?

Never Sometimes Always

Is your child involved in any form of employment?

Never Sometimes Always

Does your child accompany you or any relatives to the store or any agency and serve as a translator for you?

Never Sometimes Always

Is your child responsible for caring or for supervising his/her younger siblings while you are not at home?

Never Sometimes Always

Could your child cook/iron/go grocery shopping? Please circle the ones your child does well.

Please list any other domestic/community chores for which your child is responsible.

Implications for Psychologists

Routinely, school psychologists make diagnostic decisions without considering the possible effects of culture as mediating and intervening variables. In other words, intelligence is a multifaceted set of abilities that can be enhanced depending on the social and cultural contexts in which it has been nurtured, crystallized, and ultimately assessed. Therefore, there is a need to expand the notions of intelligence that have been developed within the psychometric and information-processing traditions. It is amazing that we live in a society so advanced in technology, but we use numbers and scores as primary bases for triage. The fact is that even among a group of children who are otherwise similar, there can be important ecological variations that contribute to their ecological development. This does not mean that information-processing theories and psychometric measures are not relevant in the field of intelligence. These theories, however, need to be expanded to include a more cultural and anthropological approach. The important point to note is that IQ is quite a labile concept and is quite responsive to a shift in context. Thus, contextual influences on more complex-type tasks inevitably cast doubt on current conceptualization of intelligence. It is wise to consider the research of others outside of the psychological IQ guild, such as anthropologists and sociologists, who bring the richness of the contextual influences into the testing situation.

The American Psychological Association (1993) has offered culturally relevant suggestions for practice for providers who work with culturally diverse families. The suggestions therein should be strongly considered. School psychologists should also familiarize themselves with the psychometric and nonpsychometric measures assessment tools that address the multifaceted nature of intelligence. The biocultural model of assessment is recommended as a comprehensive assessment system for the understanding of a qualitative, quantitative, cognitive, social, emotional, cultural, and behavioral comprehensive assessment. Although the recommended methods of administering and scoring these traditional tests may seem quite unconventional and not in keeping with standardization procedures, they are a guide for psychologists in tapping the potential of culturally different children. Reporting two IQ scores—one following standard procedures and one taking into consideration issues raised in potential assessment—may be the best way to understand these children's strengths and weaknesses. If school psychologists intend to serve children well, they should give greater attention to qualitative rather than quantitative reports that highlight all a child's past and present cultural experiences.

In summary, this "assessment view" seeks to connect school activities with after-school activities, with emphasis on the individual's strengths. In other words, this approach calls for a broader menu of assessment options and an abandonment of the sophomoric mentality that relies on some type of rigid superficial conformity. A broader trained cadre of workers would make greater use of the many subsets of human talents by embracing this assessment approach. The question that one should always keep in mind is whether the tasks on the decontextualized intelligence tests bear any resemblance to the values held by the surrounding community. Many traditional test developers still define intelligence as a unitary attribute with a cognitive overtone situated only in the individual's head. More contemporary researchers recognize that intelligence is really a flexible, mobile, culturally dependent construct. Testing in most North American contexts, however, is formalized and examines only the individual, not the individual vis-à-vis his or her community. These tests require people to examine decontextualized tasks rather than examining how people function when they can draw on their experiences and knowledge as they typically have to do in the real world. The extensive attention given to the cognitive information-processing components in testing situations is based on an assumption that the same processes are required to function in real-life contextual situations. Knowing about abstract analogies, or word meanings in isolation, does not mean that one's human intellectual performance is adequately represented. The ideal contemporary approach is one in which standardized tests are only one component of a broad-based intellectual evaluation and in which interviews with parents and observation of children in their natural setting—community, home, school, and so on—is equally valued.

All the information presented in this chapter should be incorporated in the report in a qualitative and descriptive manner (see Chapter 8 for a sample biocultural report). Table 6.3 provides a cursory view of the various biocultural measures that should be given to all involved in the assessment of a child.

TABLE 6.3 Checklist of Tests to Be Given to the Child, Parent, and Teacher

Child

- Child consent form
- Medical examination (observation of the child's school records—nutritional and physical)
- A psychometric measure or cross-battery testing
- Nonpsychometric measures
  - Item Equivalency Assessment Measure
  - Other Intelligences Assessment
  - Test-Teach-Retest Assessment Measure
  - Ecological Taxonomy of Intellectual Assessment
  - Stage of Acculturation

Parent

- Parent consent form
- Family/Community Support Assessment of linguistic proficiency, dominance, and fluency
- Medical examination (ask the parent for the child's nutritional and physical status)
- Other Intelligences Assessment
- Some questions to help assess the item equivalencies (see Family/Community Support Assessment)

Teacher and school

- Medical examination (ask the school to observe the child's medical records)
- Teacher questionnaire
- Other Intelligences Assessment

7

A Critical Review of Standardized Tests and Approaches of Intelligence Using the Biocultural Assessment System

In 1987, a nationwide survey of school psychologists was conducted by Obringer (1988) in which respondents were asked to rank in order of their usage the following instruments: the Wechsler scales, the Kaufman Assessment Battery for Children (K-ABC), and the old and new Stanford-Binets. The results were as follows: The Wechsler earned a mean rank of 2.69, K-ABC of 2.55, followed by the old Stanford-Binet of 1.98 and the new Stanford-Binet Fourth Edition of 1.26. To date, no series of intelligence tests has yet to equal the success both in practice and in research of the Wechsler scales. Despite the volumes of criticisms, the Wechsler scales continue to "enjoy unprecedented popularity and have a rich clinical and research tradition" (Kamphaus, 1993, p. 125). As such, this test will be used as the main point of reference for teaching potential assessment. In Chapter 2, we gave an overview of the conceptions of human intelligence and a cursory glance at the history of intelligence tests. A more extensive and detailed representation of the history of IQ testing can be found in French and Hale (1990) and Sattler (1988).

This chapter contains a brief description of the major scales of intelligence, their assumptions about what constitute intelligent behavior, and their limitations. An exploration of how to use the biocultural model with two of the four commonly used IQ tests will be presented in this chapter as well.

In 1905, when Binet and Simon developed the Binet-Simon Scale it became known as the first practical intelligence test (Sattler, 1988) that initially served to diagnose levels of mental retardation and was considered the prototype of subsequent measures for assessing mental ability. The age scale format was revised in 1908 and 1911, and in 1916 it was renamed by Terman, a professor at Stanford University, the Stanford-Binet. This version was revised in 1937, 1960, 1972, and 1986, when it first became a point-scale format—the Stanford-Binet Intelligence Scale: Fourth Edition (Thorndike, Hagen, & Sattler, 1986).

The development of the Wechsler scales began in 1939 when David Wechsler, a clinical psychologist at Bellevue Hospital, introduced the Wechsler-Bellevue Intelligence Scale Form I, which was the forerunner to the Wechsler Intelligence Scale for Children (WISC) (1949) and its revision—The Wechsler Intelligence Scale for Children-Revised (1974). The most daring changes of the WISC were made in the 1991 Wechsler Intelligence Scale for Children-III (WISC-III) (Kamphaus, 1992). The Wechsler scales and its derivatives—the Wechsler Primary Preschool Scale of Intelligence-Revised and the Wechsler Adult Intelligence Scale-Revised continue to enjoy much popularity and are still the most widely used tests of intelligence possibly because of their ease in administration and scoring (Detterman, 1985). Opinions on the Wechsler scales range from glowing reports to outright condemnation (Witt & Gresham, 1985).

The K-ABC (Kaufman & Kaufman, 1983) has become quite popular within the past 10 years both in the school and in clinical settings.

TABLE 7.1 Assumptions Made in IQ Tests

	Test				
Assumption	*Wechsler*	*Stanford*	*WJTCA-R*	*K-ABC*	*Biocultural*
Speed is critical	X	X	X	X	✓
Linguistic equivalence	X	X	X	✓	✓
Do problems mentally	X	X	X	X	✓
Conceptual equivalence	X	X	X	✓	✓
Nonverbal culture fair	X	X	X	X	✓
Functional equivalence	X	X	X	X	✓
Psychometric equivalence	X	X	X	X	✓
No other intelligences are needed	X	X	X	X	✓
Do not need practical or tacit knowledge	X	X	X	X	✓
IQ is fixed	X	X	X	X	✓

X, This assumption is made by this test, thus limiting the accurate assessment of the child's intelligence.
✓, This assumption is not made by this test or model.

What has made the K-ABC most appealing is its attempt to enhance fair assessment of minorities, limited English-proficient children, individuals with speech and language difficulties, bilingual children, and youngsters with learning disabilities. A strength of the K-ABC is the fact that there are "teaching items" to ensure that children understand the demands of the tasks. Although we view the K-ABC as a promising innovative test with good technical quality, like its predecessors there are still limitations with respect to its assumptions (Table 7.1).

The Woodcock-Johnson Tests of Cognitive Ability-Revised (WJTCA-R) (1989) was first introduced in 1977 as the Woodcock-Johnson Tests of Cognitive Ability—a combined psychoeducational battery. Its revised form has been touted by its followers (McGrew, 1994) as the "first intelligence battery to incorporate a number of innovations in intelligence testing" (p. xiv). As such, the limitations of this intelligence test will be presented along with a demonstration of how to conduct potential assessment vis-à-vis the biocultural model. Of course, the main problem with the WJTCA-R is that it is organized around the Gf-Gc (general/fluid-general/crystallized) theory, a theory whose comprehensive empirical-based research emphasizes that the nature of intelligence is inferred by observable differences in intellectual functioning as measured by standardized tests of intelligence. Many experts in the IQ testing industry (Ceci, 1990; Gordon, 1988; Hilliard, 1996; Sternberg, 1986) have repeatedly echoed Anastasi's (1988) position that there is "no culture fair test" and that the heavy reliance on standardized assessment measures to determine intelligence runs counter to the guidelines recommended by many experts for nondiscriminatory assessment for culturally, ethnically, and linguistically diverse students.

In general, all the IQ tests make many assumptions about intelligence that have served as the major criticisms of standardized tests of intelligence. These assumptions will be discussed in the next section.

Assumptions and Criticisms of Intelligence Tests

Table 7.1 lists several assumptions that most IQ tests make about the intelligence of human beings throughout the world. An examination of these assumptions would shed some light on the limitations and criticisms of IQ tests as a whole. All these assumptions are discussed in Chapter 6, in which the biocultural model was presented.

The first and most obvious assumption is that speed of mental functioning is a critical component of intelligence. Hence the interjection of speeded items on most IQ tests.

A second assumption is that vocabulary is only moderately influenced by culture, resulting in the insertion of vocabulary words in a decontextualized manner on many commonly used IQ tests despite the fact that most vocabulary is contextually determined and some words have different meanings in a different context (linguistic equivalence) (Hilliard, 1979).

The third assumption is that children who cannot perform mathematical computations mentally are less skilled in math and less intelligent than their counterparts who are able to do so. This has led to the inclusion of mental computations without the aid of paper and pencil in many IQ tests. Such tests do not distinguish between whether the child has the skill, whether anxiety is impeding the child's functioning, or whether the child is unable to work in a speedy manner because many of these arithmetic tests are also timed.

The fourth assumption that characterizes most IQ tests is that all children enter the assessment process with the same level of novelty or experience irrespective of their cultural backgrounds (conceptual equivalence). Thus, most IQ tests fail to recognize that some children were never exposed to some items even though they may have been exposed to the concept itself in a different form. As such, with the exception of the K-ABC that allows for teaching to ensure understanding of the tasks, most IQ tests do not accommodate mediated learning experiences.

The fifth assumption is that nonverbal tests are more culture fair. This has resulted in the development of more nonverbal-type tasks on IQ tests within the past 5 years.

The sixth assumption is deeply embedded in Westernized ethnocentric thinking whereby it is believed that Western education nurtures disembedded, context-free thinking (functional equivalence). Therefore, most IQ tests assume that one's cultural experiences and context are integral to the development of one's cognition. As such, the tests leave no room for the probability that the child may have the same skill but may not be representing it in the way the IQ test dictates. Therefore, because the child may know how to build a fan but not know how to put a puzzle together (as required on an IQ test)—two different tasks that demand a similar type of skill—the individual is not credited for having the ability because this skill or knowledge is not captured in the exact manner on the IQ test.

The seventh assumption is that the IQ test can capture the broad-based nature of a person's intelligence and therefore there is no need for a broader menu of assessment options (psychometric equivalence). Hence the reason IQ tests are used in isolation of the child's various other ecologies, such as the family and the community.

The eighth assumption is that all cognitive abilities exist within the IQ tests themselves. Hence, most psychometricians believe that children who do not perform well on the IQ tests have no other intelligences, such as musical and bodily kinesthetic intelligences.

The ninth assumption is that emotional and practical intelligence (tacit knowledge) is not a criterion for cognitive intelligence. Of course, cognitive theory has yet to address why the "most intelligent" child in the class may not always be the most successful or why some people stay buoyant when faced with great challenges when the same situations tend to sink a person who is less resilient. Not one IQ test takes into consideration emotional factors even though "the majority of variance in real-world performance is not accounted for by intelligence tests scores" (Sternberg, Wagner, Williams, & Horvath, 1995, p. 913).

The tenth assumption is that because of the genetic nature of intelligence, that intelligence is fixed and cannot be changed. This position has prohibited psychologists from functioning as diagnosticians, and instead they function as psychometricians whereby they are expected to adhere strictly to standardized procedures.

The Demise of Standardized Tests as a Sole Means of Determining Intelligence

We have begun to witness the demise of the term IQ, which has been touted as outdated by some researchers (Kamphaus, 1993). With the exception of the Wechsler scales, which have maintained an IQ score, a standard score has been given in its place by several standardized tests, such as the Stanford-Binet (composite standard score), the K-ABC (mental processing composite score), and the Woodcock (standard score and grade and age equivalent). The days of the term IQ seem to be numbered, and this is probably symptomatic of the changing climate of intelligence testing as a sole means of assessing intelligence. Given the limitations of all these IQ tests and the ongoing debate with IQ tests as a whole, Kamphaus (1993) hypothesizes that the term intelligence test may become extinct. If standardized tests are to be preserved, especially in a multiethnic and multicultural society, potential and ecological assessment must be part of the overall assessment system that we term the four-tier biocultural assessment system.

The next section presents an overview of how potential assessment can be performed with two intelligence tests.

TABLE 7.2 Correcting the Biases on the Wechsler Scales Through the Biocultural Assessment System

Subtest	*Assumption*	*Potential Assessment*
General information	Functional Equivalence	Cultural/Item Equivalence
Similarities	Functional Equivalence	Cultural/Item Equivalence
Vocabulary	Linguistic Equivalence	Linguistic Equivalence Contextualize Words
Arithmetic	Do problems mentally Speed	Paper and Pencil Suspend Time
Comprehension	Functional Equivalence	Cultural/Item Equivalence
Block Design	Conceptual Equivalence Functional Equivalence Speed	Test-Teach-Retest Cultural/Item Equivalence Suspend Time
Object Assembly	Conceptual Equivalence Functional Equivalence Speed	Test-Teach-Retest Cultural/Item Equivalence Suspend Time
Picture Completion	Speed Functional Equivalence	Suspend Time Cultural/Item Equivalence
Picture Arrangement	Conceptual Equivalence Functional Equivalence Speed	Test-Teach-Retest Cultural/Item Equivalence Suspend Time
Coding	Speed	Cultural/Item Equivalence

Administration Procedures for Potential Assessment for Two Standardized Tests of Intelligence

Wechsler Intelligence Scale for Children-III

Please follow the standardized testing procedures unless stated below (see Table 7.2 for a cursory view).

Divide your protocol in half on each subtest. Write in black or blue for actual testing scores and red for the potential testing scores. Place actual testing scores to the left and potential testing scores to the right.

If the child has reached his or her ceiling point on standardized testing but not on potential, you are to continue doing potential testing until the child has reached the ceiling point.

In potential assessment, time should be suspended on all timed tests.

Picture Completion:[1] Suspend time and let the child go beyond the "20." Note in the Response column "Time 25" and put in red in the potential column "1" if the child got the answer correct.

General Information: Match to culture. Give potential scores in red.

Coding: Match to culture. Give potential scores in red.

Similarities: Match to culture. Give potential scores in red.

Picture Arrangement (teach and suspend time):[1] Begin with the sample item as in the standardized manual. If the child passes the sample item, move along to Item 1 for children ages 6 to 8 and Item 3 for children ages 9 to 16. Items 1, 2, and 3 can be used as teaching items. Therefore, if a 9-year-old child fails Item 3, teach it, then go back and allow the child to do it again. Credit the child under potential and continue with Item 1 and 2 in normal sequence as in the standardized procedures. If the child passes Items 1 and 2, continue to Item 4, as you would have in the standardized procedure. Remember for Items 1 and 2 to give both trials. Teaching comes after the second trial. The important point to remember is that the exact procedures are followed as in the standardized testing, except time is suspended, teaching is done, and potential scores are given after the child passes the teaching items. Remember to circle the correct points (based on the timed completed) under potential scores as you would have in the standard procedure.

Arithmetic:[1] For potential testing, use paper and pencil and say to the child who fails: "Please use this paper and pencil and try to solve the problem." Circle in red the correct response because this falls under potential testing. Another version of potential is to allow the child to read all the questions as he or she would have done in Questions 17 to 19. Arithmetic taps skill, memory, attention, and speed. In the standard procedure, it is difficult to tell which is operating. In the first potential example, we can rule out speed as a factor, whereas in the second potential example, we can rule out speed, memory, and attention. Reading may be a confounding variable, however.

Block Design:[1] Begin with Design 1 for children ages 6 and 7 and Design 3 for children ages 8 to 16. If a child fails the beginning item on both trials, teach it and give the test again. Give credit under potential if the child gets it correct. Then follow standardized procedures by going to the next item in normal sequence. Therefore, for children ages 8 to 16, teach at Item 3 and give credit if the child gets it correct after the teaching. Then go back to Items 1 and 2, give credit if correct, then go to Item 4. Remember on Items 1, 2, and 3 to give both trials. Teaching comes after the second trial. The important point to remember is that the exact procedures are followed as in the standardized testing, except time is suspended, teaching is done, and potential scores are given after the child passes the teaching items. Remember, place in the incomplete design column in red the time it took, and circle the correct points (based on the timed completed) under potential scores as you would have in the standard procedure.

Vocabulary:[1] Contextualize all words by asking the child to say them in a sentence. Credit is only given if the child (not the examiner) says it in a sentence. Follow the standardized procedures for querying. Thus, if the child is on potential and he or she gives a query response, query it. If, after querying, the child gets it correct, give credit under potential. You can keep querying if the child is in a query category as in the standardized procedure, but the responses fall under potential scores.

Object Assembly:[1] Begin with the sample item for all as in the standardized procedures. Then proceed to Item 1. If the child fails Item 1, even if he or she got the sample item correct, teach Item 1 and give the credit under potential. Only the sample item and Item 1 can be used for teaching. Remember to suspend time and circle in red the potential score and give the correct points based on the timed completed.

Comprehension: Match to culture.

Digit Span: Match to culture.

Woodcock-Johnson Test of Cognitive Ability-Revised

Please follow the standardized testing procedures unless stated below (see Table 7.3 for a cursory view).

Divide your protocol in half on each subtest. Write in black or blue for actual testing scores and red for the potential testing scores. Place actual scores to the left and potential scores to the right.

If the child has reached his or her ceiling point on standardized testing but not on potential, continue doing potential testing until the child has reached the ceiling point.

In potential assessment, time should be suspended on all timed tests.

TABLE 7.3 Correcting the Biases on the Woodcock Through the Biocultural Assessment System

Subtest	*Assumption*	*Potential Assessment*
Visual matching	Speed	Suspend Time
Visual closure	Speed Conceptual Equivalence Functional Equivalence	Suspend Time Test-Teach-Retest Cultural/Item Equivalence
Picture vocabulary	Speed Conceptual Equivalence Functional Equivalence	Suspend Time Test-Teach-Retest Cultural/Item Equivalence
Analysis-synthesis	Speed	Suspend Time
Visual-auditory learning	Speed	Suspend Time
Cross out	Speed	Suspend Time
Picture recognition	Speed Functional Equivalence	Suspend Time Cultural/Item Equivalence
Oral vocabulary	Linguistic Equivalence	Linguistic Equivalence Contextualize Words
Concept formation	Speed	Suspend Time
Listening comprehension	Functional Equivalence	Cultural/Item Equivalence
Verbal analogies	Functional Equivalence	Cultural/Item Equivalence

In general, there are many tests of memory on the WJTCA-R. Therefore, this can be used to supplement the memory tests on the Wechsler scales. Also, the symbols on the WJTCA-R are unfamiliar symbols for any child from any culture. Thus, to a large extent, they are more universal symbols.

Test 3—Visual Matching:[1] Suspend time.

Test 4—Incomplete Words: Allow for different pronunciation based on different accents. If you are unsure, allow someone from the child's culture to administer this subtest.

Test 5—Visual Closure:[1] Suspend time. Also, please note that some culturally different children may not have had exposure or experience with some of these pictures (Examples include Nos. 11, 17, 18, 22-26, 28, 29, 33, 44, and 46).

Test 6—Picture Vocabulary: Please note that some culturally different children may not have had exposure or experience with some of these pictures (Examples include Nos. 6, 8, 19, 30, 35, 40, and 47).

Test 7—Analysis-Synthesis:[1] Items 26 through 35—suspend time.

Test 8—Visual-Auditory Learning:[1] Suspend time.

Test 10—Cross Out:[1] Suspend time.

Test 11—Sound Blending: Allow for different pronunciation based on different accents. If you are unsure, allow someone from the child's culture to administer this subtest.

Test 12—Picture Recognition:[1] Suspend time.

Test 13—Oral Vocabulary: (Note that even in the standard procedure, the child is not penalized for speech pronunciation differences.) On this test, you can contextualize all words by asking the child to say them in a sentence. Credit is given only if the child (not the examiner) says the words in a sentence. Follow the standardized procedures for querying. Thus, if the child is on potential and he or she gives a query response, query it. If, after querying, the child gets it correct, give credit under potential. You can keep querying if the child is in a query category as in the standardized procedure, but the response falls under potential scores.

Test 14—Concept Formation:[1] For Items 22 through 34, suspend time.

Test 18—Sound Patterns: Allow for different pronunciation based on different accents. If you are unsure, allow someone from the child's culture to administer this subtest.

Test 20—Listening Comprehension: (Note that even in the standard procedure the child is not penalized for speech pronunciation differences.) Also, please note that some culturally different children may not have had exposure or experience with some of these items (e.g., No. 3; in many Third World countries, a shower is more commonly found than a bathtub).

Test 21—Verbal Analogies: (Note that even in the standard procedure, the child is not penalized for speech pronunciation differences.) Some questions assume that the child may have been exposed to these items (Examples include Nos. 2, 8, 18, 27, and 29).

Although the assumptions and limitations of the Stanford-Binet and the K-ABC were presented in Table 7.1, a detailed subtest analysis will not be discussed here. Examiners, however, could use the previous format in developing potential assessment for any tests because all standardized tests are limited with respect to ecological assessment. Gopaul-McNicol, Elizalde-Utnick, Nahari, and Louden (1998) gave a critical review of 16 commonly used intelligence tests with bilingual children. A perusal of this series will prove helpful in understanding the limitations of other tests not discussed in this chapter, such as the McCarthy scales, the Detroit Test of Learning Aptitude 3, the Differential Ability Scales, the Leiter, the System of Multicultural Pluralistic Assessment, and the Peabody Picture Vocabulary Test and its Spanish companion, the Test de Vocabulario Imagenes Peabody.

Inter-Battery, Cross-Battery, or Process-Oriented Approach

Thomas (1990) helped to bring to the forefront the blending of IQ tests at the American Psychological Association Convention in Boston in 1990. In her discussion on how best to assess the intelligence of Hispanic children, Thomas mapped out how another aspect of the child's cognitive functioning can be ascertained by using an amalgamation of IQ tests—the Wechsler, the Stanford-Binet, the Test of Non-Verbal Intelligence, and the Kaufman. Thomas, a licensed clinical, school, and bilingual (Spanish) psychologist, personally assessed approximately 600 children during a 10-year period utilizing this interbattery approach and concluded that "it is still just another IQ test." As such, "It must be used in conjunction with nonpsychometric measures to best capture the range of a child's intelligence in various contexts and with varied experiences."

Proponents of interbattery testing, termed by other psychometricians (McGrew & Flanagan, 1995) as cross-battery testing or the process-oriented approach (neuropsychology), do not recognize that adding more items and mixing various tests in no way strengthens the validity of these tests in assessing the intelligence of children. The grave question that comes to mind with this interbattery approach to assessing intelligence is the issue of norming. Questions such as the following need to be answered:

1. Which population sample was used from each of the various tests to form this new test?
2. Why were the items selected from the various tests chosen over other items?
3. At what point is it safe to say that the examiner has exhausted all possibilities of assessing the particular concept under study? In other words, should we use three, four, or five different tests?
4. What happens if the child still fails the interbattery? Does this prove without a doubt that the child is really limited in that area being assessed? In other words, is the examiner going to conclude that it now justifies the biological explanation although item equivalence via ecological assessment was not done?

Even with this interbattery approach, the following question still remains: How can one tell that these abilities are not tainted by cultural factors? In other words, how could one know that it is ability and not experience and context that are operating? Could one say emphatically that all cultures and all people demonstrate abstract reasoning in the way they do on standardized tests of intelligence. We view this cross-battery approach as no more than a last attempt by desperate psychometricians to hold on to the cornerstone of traditional IQ tests.

Note

1. These subtests are most applicable to potential assessment.

8

Report Writing Utilizing the Four-Tier Biocultural Assessment System

Traditionally, many school psychologists consider report writing a burdensome task and tend to assume this challenge only when forced to at the end of the semester when writing the report cannot be further postponed. Despite its unpopularity, however, documented written reports are critical to the role of psychologists. First, they provide accountability and proof of the examiner's findings that have long-term effects on a child's life. Surber (1995) noted that "Probably no other profession or specialty do legislative regulations have such an important effect on the information presented in a written report" (p. 161). The purpose of the psychological report has traditionally been to address the reason the student is having difficulty in learning, behaving, and so on and to determine what services or class placement the child may need. A tremendous void in most reports is the conversion of the assessment data into specially designed tailored interventions that really fit the child's needs and that can lead to improved student performance in the areas of concern.

This chapter first demonstrates how the four-tier biocultural assessment system can assist school psychologists in writing more culturally sensitive reports instead of narrowly focused reports that are endemic to psychometricians or "tester-technicians" (Tallent, 1993) who restrict their reporting to test results only. When the psychologist's role is expanded to that of a diagnostician or a clinician, the approach to report writing is more prescriptive and relates more to educational implications for classroom practice (Gopaul-McNicol & Armour-Thomas, 1997a). Hence, the second purpose of this chapter is to assist school psychologists to develop interventions that can assist the teacher and the parent in reducing the discrepancy between the child's current functioning as assessed by classroom-type tasks and the child's potential functioning as assessed by this comprehensive biocultural assessment system. It is critical to note that in assessing students from linguistically and culturally diverse backgrounds, their personal competencies, such as their ability to negotiate an environment that is highly different from that of their culture of origin, along with their many other intelligences and personal strengths, are important in understanding the strengths of the child. Also with these children, it is necessary to conduct differential diagnoses for intellectual assessment. In other words, it is important to rule out whether the child's deficits are a result of mental retardation, educational deprivation, learning disability, a linguistic factor, or a mere misassessment of the child's strengths due to using only standardized measures of assessment.

Recommendations

After reporting all the findings, it is recommended that a Diagnostic Impression section and an Educational/Clinical Implications section—that is, what are the implications of the findings for the child's functioning in the classroom setting—be included. It is important to remember that the recommendations should be based more on the results of the intellectual potential findings, ecological findings, and other intelligences rather than only on the standardized psychometric intelligence tests scores.

In addition, the recommendations must be based on the diagnostic impressions. Thus, after doing potential intellectual assessment, the examiner should recommend what he or she believes are the best ways the teacher, the parent, and the mental health worker can intervene in working with the child. In other words, if teaching helped, then recommend one on one teaching for a particular number of sessions. If extending time helped, then recommend that the child be given extra time and more opportunity for practice. If contextualizing words helped, recommend that initially as the child acclimates to the new environment that he or she be given an opportunity to receive his or her assignment in a surrounding context. If it was found that the child did better on paper and pencil tasks than on tasks requiring mental computations, then recommend that paper and pencil assessment be allowed. If the child has other intelligences, recommend programs in which these can be further enriched.

It is also important to utilize all the resources in the community—church, social and recreational community programs, after-school programs, legal aid, psychotherapeutic programs, and so on. The important role of the psychologist is to assist the school-based support team, the teacher, the family, and the child to develop a course of treatment that would maximize every opportunity for the child to move from his or her actual functioning to his or her potential functioning in a 3-year period. In other words, the child should show significant gains after the intervention period in all areas assessed.

To best address the purpose of this chapter, a review of a traditional psychological report and a biocultural report conducted on the same child will aid the examiner in detecting how the differences in assessment and report writing can lead to differences in recommendation and improved life chances for children with special needs. The following is information on the child discussed in this chapter:

Name: Miguel	Date of testing: 5/21/94
School: JHS	Date of birth: 8/19/81
Grade: 6th	Age: 12 years, 9 months
Language: Spanish	

Reason for Referral and Background Information

Miguel was referred for an initial evaluation by his teacher due to academic difficulties. Miguel arrived from Colombia in August of 1993 at age 12. His parents had migrated to the United States when he was 7 years old. While in Colombia, Miguel was said to be a pleasant boy who related well to his aunts with whom he resided while his parents were in the United States.

The social history conducted in March 1994 by the school social worker revealed that since Miguel arrived in the United States, he has had serious difficulty adjusting to the classroom not because of linguistic factors "Because he is more proficient and more dominant in English since he attended English classes in his native country to prepare for his migrating to the USA." According to the social worker and the teacher, Miguel clearly has no skills, and the school psychologist's report revealed deficient intellectual functioning when tested with standardized tests of intelligence.

The social history revealed that Miguel lives with his mother, father, an older brother, maternal aunt, and grandmother. All family members present themselves as a cohesive unit with strong extended family ties and good family support systems. According to Miguel's mother, all developmental milestones were attained at age-expectant levels. There were reports of delays in reading even in Colombia, however, but his mother said, "He was certainly able to read what was necessary to get by. He is definitely not stupid as the school is making him out to be."

Test Administered and Test Results: Traditional School Report

Wechsler Intelligence Scale for Children-III

Psychometric Assessment	*Range*
Verbal Scale IQ	Deficient
Performance Scale IQ	Borderline
Full-Scale IQ	Deficient

Wechsler Intelligence Scale for Children-III (continued)

Current Scale Score	*Range*	*Current Scale Score*	*Range*
Information	Deficient	Picture Completion	Borderline
Similarities	Deficient	Coding	Borderline
Arithmetic	Deficient	Picture Arrangement	Borderline
Vocabulary	Deficient	Block Design	Borderline
Comprehension	Deficient	Object Assembly	Borderline
Digit Span	Low average	Mazes	Borderline

Given the previous findings, the standardized psychological report reflected that Miguel is functioning in the mentally deficient range of intelligence, and even when tested to potential (extending the time limits in some of the nonverbal areas), he was at best borderline. He was unable to do basic math commensurate to his age and grade peers. He was deficient in his vocabulary skills, and could not define words such as nonsense, ancient, and thief. When given the blocks and puzzles to manipulate, he became noticeably frustrated, stating "I do not know what to do," while pushing the blocks away. On the general information subtests, Miguel was unable to respond to basic questions such as "Name two kinds of coins." The report recommends that Miguel be placed in a self-contained class in which he can receive special educational services in all areas to address his obvious delays in academic-type tasks and his overall intellectual deficiencies.

Test Administered and Test Results: Biocultural Assessment System

Wechsler Intelligence Scale for Children-III

Psychometric Potential Assessment	*Range*
Verbal Scale IQ	Borderline
Performance Scale IQ	Average
Full-Scale IQ	Low average

Potential Scale Score	Range	Potential Scale Score	Range
Information	Deficient	Picture Completion	Low average
Similarities	Deficient	Coding	Borderline
Arithmetic	Low average	Picture Arrangement	Low average
Vocabulary	Average	Block Design	Average
Comprehension	Borderline	Object Assembly	Average
Digit Span	Average	Mazes	—

Medical examination: No medical difficulties

Ecological Intellectual Assessment	
Estimated overall functioning	Average
Other intelligences assessment	
Bodily kinesthetic (soccer)	Advanced
Artistic (painting)	Advanced
Musical intelligence (guitar)	Advanced
Family Support Assessment	Adequate

Vineland Behavior Adaptive Scales—Parent edition	Range
Communication	Low
Social	Adequate
Daily Living scales	Adequate
Social History	
Clinical Interview	
Parent Interview	
Teacher Questionnaire	
Language Dominance	English
Language Proficiency	

Vocabulary Subtest	Proficiency Rating
Spanish Expressive Vocabulary	Deficient
English Expressive Vocabulary	Borderline

Behavioral Observation

Miguel, a pleasant young man, presented himself in a cooperative, compliant manner. He had a good disposition and was motivated to do all the tasks assigned to him. Even on completion of the testing, Miguel asked the examiner if he could do more. He was not fatigued and believed these types of tests were reinforcing to him. In general, his response time was slow, and he approached the testing in a cautious, reflective manner. When he clearly did not know the answer, he still persisted but became noticeably frustrated and embarrassed. He would sigh, frown, and seemed upset that he did not know the answer to a question that he initially perceived as easy. All in all, it was a pleasure testing Miguel because he tried hard and was willing to please.

Language Assessment

Miguel's language proficiency was tested through the administration of the vocabulary subtest of the Wechsler Intelligence Scale for Children-III (WISC-III) in both Spanish and English. He is clearly more dominant and more proficient in English. He spoke mainly in English, but on several occasions he requested that the examiner speak in Spanish. When he engaged in social play on the playground, he spoke in both languages. In general, his English receptive and expressive skills are better developed than his Spanish skills. It is possible that the years of English instruction he received prior to coming to the United States and the fact that his parents speak only in English at home aided in his developing such proficiency and fluency in English. At this time, although Miguel can function well in a predominantly English-speaking class, the supportive environment of a bilingual paraprofessional may prove beneficial when he is faced with very difficult tasks.

Test Interpretation

Psychometric Assessment

On the WISC-III, Miguel obtained a full-scale IQ score that placed him in the deficient range of intelligence. His verbal and nonverbal scores fell in the deficient and borderline ranges, respectively.

Subtest analysis indicates considerable subtest variability within both the verbal and nonverbal spheres. In the verbal area (crystallized), Miguel was deficient in general information, suggesting that on this psychometric test, Miguel is not as alert to the social and cultural factors typical of American society as measured by the WISC-III. His deficiency in comprehension is also indicative of his limited understanding of the social mores in the United States as assessed on the WISC-III. Miguel was also deficient in verbal abstract reasoning, suggesting that, on this test, Miguel has difficulty placing objects and events together in a meaningful group. In arithmetic and vocabulary, Miguel was also deficient. This is indicative of inadequate arithmetic skills as assessed by the standardized procedures of the WISC-III as well as poor language development and limited word knowledge as defined by the WISC-III. In auditory short-term memory, Miguel was low average. Therefore, one can expect Miguel to be relatively good at rote memory and sequential processing.

In the nonverbal area (fluid intelligence), Miguel was borderline in identifying essential missing elements from a whole, suggesting delayed visual alertness, visual discrimination, and long-term visual memory on the Wechsler scales. In visual integration, Miguel was borderline, suggesting limited perceptual skills, poor long-term visual memory, and limited constructive ability commensurate to his peers nationwide on the Wechsler scales. Miguel, however, was persistent and tried to put the puzzles together. There was a sense that he was unfamiliar with these items. As such, when he was taught how to connect the pieces, he tended to be more relaxed, although he continued to perform poorly. In visual motor coordination and motor speed, Miguel was also borderline, suggesting slow response time, poor visual short-term memory, and limited visual acuity on the Wechsler scales. In nonverbal comprehension, Miguel was borderline, suggesting a delayed ability to anticipate the consequences of his actions, to plan, and to organize ahead of time. In nonverbal abstract reasoning, Miguel was also low average, suggesting below average ability to perceive, analyze, and synthesize blocks on the Wechsler scales.

Psychometric Potential Assessment

When Miguel was tested to the limits such as when time was suspended, and also when item equivalencies were done, his scores improved by 14 IQ points in the verbal area, 16 IQ points in the nonverbal area, and 16 IQ points in overall intelligence. Thus, when the test-teach-test technique was implemented and when time was suspended with the blocks and puzzles, Miguel went from borderline to average, displaying much confidence on these tests during potential assessment. Because Block Design is the best measure of nonverbal intelligence on the Wechsler scales, Miguel is of average potential in the nonverbal area.

Another important fact is that when Miguel was offered the opportunity to use paper and pencil, he was able to perform many of the mathematical tasks that presented difficulty under standardized procedures. For instance, he clearly knew multiplication, division, and even simple fractions. Thus, by allowing Miguel to use paper and pencil, instead of relying on mental computations only, the examiner was able to determine that Miguel did master some arithmetic skills but was unable to perform them without the aid of paper and pencil. Because in real-life situations one is usually allowed the opportunity to work with pencil and paper, one can expect that Miguel will be able to do basic calculations to function adequately well in his day to day duties. Moreover, although Miguel was unable to name two American coins on the psychometric test, he was able to correctly name and identify the *escudo* and the *peso,* two monetary units from South America. Furthermore, when the vocabulary words were contextually determined—that is, Miguel was asked to say the words in a surrounding context (Miguel said, "I migrated to the USA recently," even though he did not know the word "migrate" in isolation)—Miguel score rose from deficient to average. He knew almost every word commensurate to his age and grade peers when allowed to contextualize them. Because vocabulary is the best measure of general intelligence, Miguel is of average potential in the verbal area. Incidentally, if the substitute test was used to tabulate his verbal IQ score instead of general information (the most biased of the verbal subtests), Miguel's verbal score would have been low average, albeit his overall IQ score would have still been low average.

Ecological Assessment

At home, in school, on the playground, and in the community, Miguel is described as "bright and promising" by his family and friends. According to his mother, Miguel helps with the groceries and assists with basic household tasks commensurate to those of his peers.

Moreover, in observing Miguel on the community playground, it was clearly evident that he was able to perform several of the tasks found on the IQ test. For instance, although he was unable to put the puzzles and blocks together on the Wechsler scales, he was adept at fixing a car. On one occasion, when his aunt's car was unable to start, he checked the carburetor and other car parts and deciphered the problem. His aunt mentioned that he is responsible for repairing any electrical appliances that malfunction in the home. Evidently, this activity involves the same visual motor coordination skills as putting puzzles together. The fact that Miguel was unable to reintegrate the pieces of puzzles on the IQ test, but could have assembled smaller, more complex parts of a car, suggests that cultural factors must be impeding his ability to perform such a similar task on the standardized IQ test. Clearly, he is at least average in visual motor integration skills, albeit this was not evident on the psychometric measure. Also significant was Miguel's ability to remember a 14-item grocery list, although he was unable to recall as many as seven numbers on the Digit Span subtest of the Wechsler scales. Equally impressive is his ability to do arithmetic computations mentally at the grocery store, although he demonstrated deficient mathematical skills on the psychometric IQ test. Thus, in Miguel's ecology—that is, in a real-life situation away from the testing environment—he showed good planning ability, good perceptual organization, good mathematical skills, and good short-term memory. Unfortunately, none of these skills were manifested on the standardized traditional IQ test, albeit gains were noted when he was tested for his cognitive potential via the same IQ measure. Evidently, from an ecological perspective, in real-life situations Miguel's cognitive ability is approximately average.

Other Intelligences Assessment

Miguel's family and his gym teacher described him as "multi-talented." He is said to be very athletic, particularly in his ability to play soccer. His gym teacher described him as well coordinated. He is artistic in that he paints and draws all sorts of abstract images as well as cartoon-like figures. His art teacher described his skills as advanced in artistic ability and said that Miguel was the best student in his class in all artistic-related fields, such as painting, designing, architecture, and so on. Thus, with respect to his bodily kinesthetic ability, Miguel was above average to superior commensurate to his peers. Another intelligence that Miguel possesses is his musical ability. He plays the clarinet and formulates melodic and harmonic images with fluency after only 1 year of playing the guitar. His mother stated that he also has an interest in other musical instruments such as the flute. An interview with his music teacher revealed that Miguel plays the piano "for fun" and composes songs and music so creatively that in the realm of musical intelligence he would be considered superior intellectually.

An interview with the after-school community director revealed that he is a "well-rounded, talented" young man who manifests accuracy, grace, speed, power, and great team spirit in all artistic and sports-like endeavors. He is said to have a well-developed sense of timing, coordination, and rhythm when it pertains to playing music. Also reported by the director is his ability to remain composed under great pressure. An observation of him playing soccer allowed the examiner the opportunity to note his bodily intelligence in its purest form with much flexibility and high technical proficiency. He is indeed of superior ability in the area of gross and fine motor motions. Also of note is the social feedback offered by the director: "Everyone likes Miguel, both young children and his peers. Everyone wants him to lead the team. He inspires his peers to do their best." Interpersonal skills are described as excellent because Miguel is a "warm, pleasant, and sociable young man."

Diagnostic Impression and Educational and Clinical Implications

Intellectually, Miguel is functioning in the mentally deficient range on the WISC-III psychometric test and low average range on psychometric potential assessment. Because Miguel attended school in his native country on a regular basis, he cannot be said to be educationally deprived. Also, a diagnosis of mental retardation cannot be given because adequate functioning was noted on two sections of the Vineland Adaptive Behavior Scales. To be diagnosed as mentally retarded, low functioning in all areas of social adaptation should be evident. It is only on communication skills that he was found to be low, which was commensurate to his score on the WISC-III psychometric test. Besides, after conducting a family assessment, it is clear that Miguel functions adequately in the his community and is respected by his peers. Thus, despite communication delays, there are no overall social adaptive deficiencies to characterize him as mentally deficient. At this juncture, Miguel's intellectual functioning best fits the category of Learning Disabled Not Otherwise Specified. This category is for learning disorders that do not meet the criteria for any specific learning disorder, and it may include problems in all three core areas of reading, mathematics, and written expression.

Given the obvious delays in several academic skill areas, and because of the psychometric IQ test scores, one (such as the school psychologist) may be solely inclined to provide Miguel with intensive instruction in all academic cognitive skill areas on a daily basis in a small classroom special educational setting. Clearly, he does require the supportive environment of supplemental instruction. Given his performance when assessed in other settings beyond the IQ testing environment, however, a less restrictive setting outside of the special education self-contained realm should be explored. For instance, Miguel should be encouraged to pursue music, in particular the guitar. Likewise, he should be encouraged to embellish his athletic skills given his intellectual prowess in this area. As such, the typical special education self-contained class, in which there is little emphasis on honing one's career and occupational skills, is not recommended.

Evidently, Miguel's obvious intelligence in music renders him a prime candidate for a scholarship at a music school. As such, opportunities for career-related academic skill development, which includes essential work-adjustment skills, and direct work experience through daily practice in a music school are needed for this young man to attain his potential and be self-supportive.

Clinically, Miguel was lacking in self-confidence and was noticeably frustrated when faced with demanding classroom-type tasks.

When observed in the home and in his music and sports classes, however, no frustration nor anger were noted. Likewise, the after-school instructor reported that Miguel had a good sense of himself—the opposite to what his general classroom teacher had stated. It was indeed important that assessment of this youngster in various settings was done to ensure a more accurate diagnosis of his overall functioning.

Summary and Recommendations

Miguel is a 12-year-old young man who showed delays on both the psychometric and the potential psychometric assessment measures in general information, comprehension, arithmetic, and verbal abstract reasoning. As a result, remediation should focus on exposure to a broad range of everyday facts and practical reasoning in social situations. Miguel should be encouraged to read American literature or the newspaper on a daily basis to gain more insight into world events and the mainstream cultural views to help improve his general information. Other forms of resources are museums, educational television shows, tapes, and film documentaries. Teaching same-different concepts should aid in improving verbal abstract reasoning. In addition, vocabulary skills can be enhanced by encouraging Miguel to learn new words by reading more. Teaching computational skills commensurate to his grade peers should aid in improving arithmetic skills.

Miguel can function in a monolingual class, although the supportive environment of a bilingual paraprofessional may prove beneficial when he is faced with difficult tasks. In addition, the recommendation for Miguel also included a referral to Operation Athlete, an organization in New York City that provides scholarships for gifted athletes. This organization has an after-school program whose goal is to recruit intelligent athletes who can go on and become professionals in their areas of expertise. Miguel was recently offered a scholarship for soccer, but he must finish high school while maintaining passing grades in all core courses.

Miguel was also referred to Sesame Flyer, a Caribbean organization that teaches immigrant families to play various musical instruments.

Moreover, counseling aimed at increasing frustration tolerance surrounding his academic delays was offered for 8 weeks.

A follow-up of Miguel's progress 1 year after the completion of the evaluation revealed a continued superiority in the nonacademic-type tasks such as sports and a slight increment in the academic areas. Miguel was taught to transfer his knowledge from his ecology to the classroom setting by various exercises offered by the examiner who continued treatment following the evaluation. Teacher and family consultation to assist those who work more closely with Miguel was offered on an ongoing basis. The most recent teacher report revealed "significant gains in reading, math, and spelling," and no frustration was evident when faced with difficult tasks. On the contrary, Miguel repeatedly stated, "This is my weak area, but I have many strengths." Miguel has learned to rely on his other intelligences—bodily kinesthetic and music—and hopes to pursue one of these vocational arenas.

Miguel should be monitored closely and tested next year to see what progress he is making and if a more or less restrictive setting would be beneficial.

Conclusion

The linking of assessment and diagnosis to intervention continues to be one of the most challenging demands faced by school psychologists. The important factor to note at this juncture is that assessment programs that fail to take into account the differences among individuals' cultural experiences are anachronistic. To take these variations into account, it would require those in the formal testing enterprise to suspend some of the major assumptions of standardized testing, such as uniformity of individuals' experiences and the penchant for one type of cost-efficient instrument. Surber (1995) emphasizes that incorporating a multimethod, multitrait approach to assessment and intervention can better ensure that the outcomes for students are nondiscriminatory. He suggests conducting more comprehensive assessment and writing more integrated reports. Indeed, such an approach allows the examiner to answer more readily the major question—What is interfering with the child's ability to learn? Such an approach expands the psychologist's role beyond that of a psychometrician who administers only standardized tests.

Psychologists in training should be taught about individual differences by being introduced formally to such distinctions. It is going to be quite difficult for students in training to arrive at such empirically valid taxonomies of differences in individuals on their own. Such exposure should occur during their professional training. Once exposed to different profiles in the course of their apprenticeships, it is easier for them to be more flexible in their assessment practices. Chapter 10 discusses issues in the training of psychologists.

Furthermore, it is equally important for students in training to be cognizant of their individual state regulations regarding bilingual and bicultural assessment. In several states, there are Chancellor's disclaimant statements for assessing bilingual children. The reader should see Table 6.2, which presents a summary of the stages in the biocultural assessment system in a biocultural report. More examples of how to word biocultural sections of a report can be found in Table 8.1. A sample disclaimant statement for bilingual children is also provided in Table 8.1.

In summary, the most critical dimension to assessment is getting at the strengths of a youngster and helping that child to feel a sense of empowerment and success despite any obvious academic deficiencies.

Although these recommended methods of administering and scoring traditional tests are not in keeping with standardization procedures, they certainly can assist psychologists in tapping the potential of all children. If psychologists intend to serve children well, they should focus on qualitative rather than quantitative reports that highlight all of a child's past and present cultural experiences.

TABLE 8.1 Examples of How to Write Sections of a Biocultural Report

Psychometric Assessment

For those who opt to interpret this test via the Gf-Gc factors, write a paragraph on each of the nine abilities (McGrew, 1994). For example:

Processing speed (GS) (coding) was found to be borderline. This suggests that Miguel may experience difficulty performing automatic intellectual tasks quickly.

Visual processing (GV) (block design and object assembly) was average. This suggests that Miguel has the ability to analyze, synthesize, and think with visual patterns. His strength clearly is his ability to manipulate visual shapes, especially those that are figural and geometric in nature. Therefore, it is not surprising that Miguel demonstrates exceptional talent in painting.

Psychometric Potential Assessment

Example 1

The Block Design and Object Assembly subtests are highly influenced by the American culture. When Miguel was assessed by other more comparable measures, such as building a chair, he was found to be very superior given the quick and accurate manner in which he executed the task.

Example 2

Miguel's scores were elevated when he was asked questions more related to his cultural experiences. For instance, although he did not know who Anne Frank was on the IQ tests, he knew Rafael Pombo—the author of the Colombian national anthem. Clearly, Miguel has some general fund of information endemic to his culture of origin. One could expect that with time, he will equally master the general type of knowledge more specific to the American cultural experiences.

Example 3

Although Miguel did not know who discovered America or who was Christopher Columbus, he knew that Pedro De Valdivia founded Santiago. Thus, if given time to acclimate to this society, it is expected that Miguel will learn American history and concepts, resulting in a higher intellectual functioning.

Example 4

On the Block Design subtest, Miguel got the more difficult items correct after he passed his ceiling point or after time limits had been expended. In his native country, blocks and puzzles are not games commonly played by children. As such, he was never exposed to block building. It seemed as if he was learning as he went along, and that lack of familiarity and ultimately anxiety may have been why he did not do as well on the earlier items. As a result, two IQs were tabulated: one following standardization procedures and one tapping his

(continued)

TABLE 8.1 (continued)

potential as evidenced by summing all points attained even after he had reached his point of discontinuation.

Example 5

Miguel was asked, "In what way are an apple and a banana alike?" and he did not know the answer. Interestingly, however, when he was asked how a mango and a banana are alike, he gave a correct answer. The important thing here is that he knew the concept of fruits.

Example 6

Although Miguel did not know the word "migrate" in isolation, he knew it when used in a sentence. Thus, if contextualized, Miguel's word knowledge and ability to express verbal ideas at varying degrees of abstraction was much higher. In the same manner, although Miguel was unable to articulate how a wheel and a ball were alike, he was able to draw how they were both alike and then to state that they were both round. It seems as if he had to first conceptualize their similarity via visual stimuli and then form the concept of sameness before being able to express the common factor between these two objects. Therefore, it is not that Miguel has not conceptualized this relationship but rather that he has to go through a longer process to retrieve and articulate the similarity of such a concept.

Example 7

Miguel was able to respond to questions previously misunderstood or unanswered when Spanish dialectical terminologies were utilized. Thus, although he did not understand the question, "Why do you recycle paper?" he was able to respond appropriately to "Why do you separate paper in a separate garbage can?" Likewise, although Miguel did not understand the question, "Why do games have rules?" he clearly was able to produce a two-point response when asked, "Why are rules needed to play marbles?" Thus, merely rewording the directions helped with understanding and resulted in an increased performance of 10 points.

Example 8

In observing (Sternberg's) atomization information process approach (Triarchic Model), when given the opportunity to slowly process nonverbal reasoning tasks by trial and error, the ability to correctly complete the tasks became successively more automatic.

Example 9

When standardization procedures were not followed, Miguel's potential demonstrated an 8-point difference. For instance, instead of presenting the visual stimuli for 5 seconds as in the standard procedure, the gestalt was shown for 10 seconds (15 seconds for more complicated gestalts), and word or number sentences were slowly repeated. This process allowed for, and resulted in, the ability to process information, increase concentration, decrease anxiety, and

TABLE 8.1 (continued)

correctly respond. This suggests that when given more time to process and practice learned problem-solving skills, he can perform quite better.

Example 10

Although initially Miguel did not understand the Block Design subtest, when a test-teach-retest technique was used (i.e., when he was taught how to build blocks and puzzles other than those used on the Wechsler scales and was then tested again), he was able to correctly synthesize the more difficult items and to correctly strategize and respond to two of three previously incorrect questions. Thus, with respect to nonverbal abstract reasoning, he is more average than borderline, as was the result when standard procedures were followed. This also shows that Miguel is capable of learning various tasks once they are explained and he is given the opportunity to practice the tasks.

Example 11

When Miguel was tested to the limits—for instance, when he was not placed under time pressure—and when item equivalencies as well as the test-teach-test techniques were implemented, Miguel's score went from borderline to low average/average in the verbal area, borderline to average in the nonverbal area, and borderline to low average/average in overall intelligence. When Miguel was asked to perform comparable skills to the puzzles on the Wechsler scales, he went from borderline to average. Likewise, when the vocabulary words were contextually determined—that is, the words were structured in a surrounding context ("Miguel migrated to the USA recently") instead of asking for a definition of the word migrate in isolation—Miguel went from low average to average. Because vocabulary is the best measure of general intelligence, Miguel is of average potential in the verbal area. In the nonverbal area, Miguel went from deficient to low average on the Block Design subtest when he was taught (test-teach-retest) to manipulate the sample block. Thus, Miguel is of low average potential in the nonverbal area because Block Design is the best measure of nonverbal intelligence on the Wechsler scales. Overall, Miguel's potential intellectual functioning is borderline.

Ecological Assessment

Observation of the child in various settings, such as on the playground or in his or her community, offers interesting information.

Example 1

Although Miguel had difficulty putting the puzzles together on the WISC-III, at home Miguel had no difficulty dismantling a fan and putting it back together in a 1-hour period. Thus, when he was exposed to a more familiar stimulus, he was able to integrate parts into a meaningful whole, which is very characteristic of the Object Assembly subtest of the Wechsler scales.

(continued)

TABLE 8.1 (continued)

Example 2

The television and Nintendo were recently broken, and Miguel was able to disconnect the wires and repair these appliances. This task is equivalent to tasks on the IQ tests, such as building puzzles and attending to detail, that require the same analysis and synthesis and visual stimulation as repairing the television and the Nintendo.

Example 3

Miguel showed the examiner a table made from logs that he assisted his neighbor in building. This task is equivalent to that of building blocks on the IQ test.

Other Intelligences Assessment

Example 1: Musical intelligence

Despite Miguel's deficiencies in the verbal area, he is able to formulate melodic, rhythmic, and harmonic images into elaborate ideas although he never studied music. For instance, he plays the steel pan, the piano, and the cuatro without sheet music and with fluency. He also composes music so creatively that in the realm of musical intelligence he would be considered superior intellectually.

Example 2

According to Miguel's parents, he is very musical and plays the guitar for various Hispanic events. His father says that he has a desire to compose music and he does it so creatively that in the realm of musical intelligence he would be considered superior intellectually.

Example 3: Bodily kinesthetic intelligence

Miguel is able to dance energetically. His dances allow one an opportunity to observe his bodily intelligence in its purest form with flexibility and high technical proficiency. He is indeed of superior ability in this area.

Example 4

Miguel is athletic and is able to excel in grace, power, speed, accuracy, and teamwork. His ability to pitch the ball shows his analytic power and resourcefulness. Also noted was his ability to remain poised under great pressure. A well-developed sense of timing, coordination, and rhythm result in his being well executed and powerful in his gross and fine motor motions. His bodily intellectual strength is indeed superior.

Example 5

Miguel has an adequate amount of social competence to deal with issues in his community. For instance, he knows which areas in his neighborhood are drug infested and how to avoid going to those areas. He repeatedly said, "Here is where the drug people hang out, so don't go there." He also knew that it was

TABLE 8.1 (continued)

unsafe to flash money around and cautioned the examiner about opening her wallet even in the supermarket.

Example 6

Miguel displayed great leadership skills in the school yard. He initiated the organization of games with ease and confidence. He delegated responsibility so that the work was evenly divided among the team players and examined the regulations of the game. Miguel was particularly sensitive to his peers' feelings, ensuring that everyone was included.

Writing a Language Assessment section

Miguel's receptive skills are stronger than his expressive skills because he had more difficulty expressing verbal ideas than understanding what was said to him. Receptively and expressively, he is more dominant in Spanish because he only responded in Spanish, even when the examiner spoke to him in English. Of note is that Miguel is also more proficient in English because he was better able to perform mathematical computations in English and to read and write in English. He counted up to three in Spanish and knew some of the letters in Spanish. In English, however, he was able to construct sentences and to do applied mathematical problems. It is possible that because he receives instruction in the classroom only in English, his English skills far surpass his Spanish skills despite being more socially dominant in his native language. The fact that the Spanish translation improved his score in all verbal areas indicates that Miguel requires the supportive environment of bilingual instruction and should be evaluated bilingually when tested psychologically, educationally, and linguistically.

Given the fact that Miguel is bilingual and that a Latino or Hispanic population was not used as part of the standardization sample, in keeping with the Chancellor's regulations, the scores should be interpreted with caution and should be used only as a guide for school personnel. The results should be interpreted from both a biological and a contextual approach.

A Disclaimant Statement for
Bilingual and Bicultural Children

At the end of the behavioral observation section, it is suggested by state officials that the following disclaimant statement for bilingual and bicultural children be included:

Because this test was not standardized on an Asian/Latino/Caribbean etc. population, the following scores should be interpreted with caution in keeping with the Chancellor's regulations in assessing bilingual and bicultural children. Thus, these scores should only be used solely as a guideline in assisting school personnel in designing the best program for this child.

Test Results: Scores

Many school districts do not permit scores to be placed in the body of the report. Miguel's scores are listed here for the reader to review the changes in scores from psychometric to potential assessment.

Wechsler Intelligence Scale for Children-III

Psychometric Assessment	*Scale Score*	*Range*	
Verbal Scale IQ	62	Deficient	
Performance Scale IQ	74	Borderline	
Full-Scale IQ	65	Deficient	

Psychometric Assessment	*Scale Score*	*Psychometric Assessment*	*Scale Score*
Information	3	Picture Completion	6
Similarities	2	Coding	6
Arithmetic	4	Picture Arrangement	6
Vocabulary	4	Block Design	5
Comprehension	3	Object Assembly	6
Digit Span	8	Mazes	6

Psychometric Potential Assessment		*Range*
Verbal Scale IQ	76	Borderline
Performance Scale IQ	90	Average
Full-Scale IQ	81	Low Average

Potential	*Scale Score*	*Potential*	*Scale Score*
Information	3	Picture Completion	8
Similarities	4	Coding	6
Arithmetic	7	Picture Arrangement	8
Vocabulary	10	Block Design	9
Comprehension	5	Object Assembly	11
Digit Span	9	Mazes	8

9

Evaluation of the Biocultural Assessment System

In Chapter 4, we proposed a biocultural perspective of intelligence in which we have argued that behaviors observed as "intelligent" should be more appropriately termed culturally dependent cognitions because, to a large extent, we think they reflect the outcomes of person-environmental interactions developed over time in particular cultural niches. For the past 10 years, we have sought confirmation of this basic thesis through our work at the Multicultural Educational and Psychological Services Agency. Toward this end, we developed the Four-Tier Assessment System of Intelligence to seek answers to the following four questions about children's intellectual functioning:

1. What is the nature of the actual cognitive strengths and weaknesses as measured by a traditional psychometric measure of intelligence?

2. What is the nature of the potential cognitive strengths and weaknesses not observed on the IQ measure?
3. In what context and for what types of experiences other than the formal testing environment are cognitive strengths displayed?
4. Do children demonstrate other intelligences beyond those that seem to be measured on the IQ test?

In this chapter, we describe the results of our work to date. We begin with the examination of data using a widely used measure of intelligence: The Wechsler scales. This is followed by a presentation of findings using procedures for cognitive potential elicitation during the actual administration of the psychometric IQ measure. Next, we present data of cognitive competencies obtained from our observations of children in contexts beyond the testing environment. In addition, we provide data on other intellectual competencies from reports of parents, teachers, and the children themselves. Finally, we end with other self-reported data from workshops and seminars conducted with practitioners and psychologists in training.

Psychometric Data

Studies From the Research Literature on Intelligence

Psychometric data were derived from three sources: factor analytic and other studies as well as our own assessment of children in two broad age groupings (6-11 and 12-16). Our review of the literature indicated that the Wechsler scales are among the more commonly used measures of intelligence in the field. Review of the factor analytic studies indicated that there was substantial evidence for a number of verbal and nonverbal cognitive abilities underlying the tasks on this measure (e.g., Reynolds, 1990; Sattler, 1988). From our biocultural perspective, we define these competencies as culturally dependent cognitions because all the subtests, to some degree, measure previously acquired knowledge and cognitive skills developed from particular experiences in particular cultural niches. There is also some support for this judgment in the intelligence test literature, although the theoretical lens through which the judgment is made is different from ours. Consider, for example, the following view of Kaufman (1979), another well-known intelligence test developer:

> The WISC-R [Wechsler Intelligence Scale for Children-Revised] subtests measure what the individual has learned. . . . From this vantage point, the intelligence test is really a kind of achievement test . . . a measure of past accomplishments that is predictive of success in traditional school subjects. When intelligence tests are regarded as measures of prior learning, the issue of heredity versus environment becomes irrelevant. Since learning occurs within a culture, intelligence tests obviously must be considered to be culture-loaded—a concept that is different from culture biased. (pp. 12-13)

More direct evidence for our conception of intelligence as culturally dependent cognitions comes from the work of Ribeiro (1980), who analyzed the data from the WISC-R administered to 350 low-income Portuguese-speaking children in Massachusetts. Ribeiro, himself a Portuguese immigrant who came to the United States as an adult, attributed the low scores obtained by these children to cultural differences in the experiences of the children. Similar findings of culture saturation have been reported by others who seek to assess children's intelligence using the Wechsler scales (e.g., Mercer, 1979).

The WISC-III or the EIWN-R (Spanish WISC-R), depending on the children's level of proficiency in English or Spanish, were administered to 244 children, 140 of whom were ages 6 to 11 and 104 ages 12 to 16. These measures were also administered to another sample of 51 children with ages also ranging between 6 and 16. Proficiency was determined by the Family Support Assessment questionnaire and the New York State School/Home Language Survey. Due to academic, behavioral, or adjustment difficulties, these children were referred for psychological evaluation by their parents or by the teachers of various New York state school districts to determine the best program placement. Table 9.1 presents a breakdown of the background characteristics of these children.

Table 9.1 Background Characteristics of Children

	Gender		*Ethnicity**			*School Type*			
Sample (Grade)	*Boys*	*Girls*	*SP-C*	*Eng-C*	*AF-A*	*Kindergarten*	*Elementary*	*Middle*	*High*
6-11	90	51	135	6	0	6	135		
12-16	71	33	95	4	5	0	27	48	13

	Language						*Program Type*				
	Dominant Language			*Proficient Language*							
	Spanish	*English*	*West Indian Creole*	*Spanish*	*English*	*Bilingual*	*Instructional Services-1*	*Supplemental Instructional Services-1*	*Other*	*Special Education*	*Regular Education*
6-11	34	82	1	26	102	13	29	8	24		80
12-16	32	64	3	23	78	3	28	9	25		42

	Age										
	6	*7*	*8*	*9*	*10*	*11*	*12*	*13*	*14*	*15*	*16*
6-11	15	20	22	22	28	33					
12-16							21	29	32	21	1

NOTE: *SP-C = Spanish Caribbean Eng-C = English Caribbean; AF-A = African American

Psychometric Potential Data

As described in Chapter 6, the basis for the psychometric potential procedures was initially derived from our clinical observations during standardized administration of the Wechsler scales as well as our postadministration interviews with children and their families. To obtain more direct evidence for our hypothesis that there was more to the children's cognitive functioning than was possible to measure from the standardized administration of the Wechsler scales, we used a number of complementary procedures (contextualization, paper and pencil, teach-test-retest, and suspending time) during the actual administration of the IQ test. We assessed two groups of children (ages 6-11 and 12-16) at two different points in time (1991 and 1992) and found far more substantial diagnostic information from the psychometric potential procedures than from the use of the IQ measure alone. A description of the samples and a comparison of the data for the psychometric and psychometric potential are found in Tables 9.2 and 9.3, respectively.

The data showed wide variation in performance between the standard and potential scores (full-scale, verbal, and performance tests). Statistical tests revealed significant differences for each age cohort (ages 6-11 and 12-16) between standard and potential scores. These findings lend support to the validity of the potential procedures to elicit information about cognitive functioning beyond what was obtained under standardized administration.

Ecological Data

Using the data obtained from the psychometric and psychometric potential procedures for the second sample of 51 children (Table 9.4), we sought firsthand evidence of the child's cognitive functioning in settings other than the testing context. We observed the child in the home, the playground, and, in some cases, the community (e.g., grocery store and the video game parlors) and used the ecological taxonomy to record the evidence. From these ecologies, we gathered a great deal of qualitative information about the children's cognitive competencies as they interacted with peers and adults in these noncontrived contexts.

TABLE 9.2 Standard and Potential Profiles of Children Aged 6 to 11

Range of Intellectual Functioning	*Score*					
	Full Scale		*Verbal*		*Performance*	
	Standard	*Potential*	*Standard*	*Potential*	*Standard*	*Potential*
Mentally retarded	6	4	14	6	2	2
Deficient	9	6	21	9	7	4
Borderline	6	15	34	30	12	9
Low average	39	48	33	39	40	28
Average	45	60	37	56	62	78
Above average	4	5	1	0	11	11
Superior	1	2	0	0	3	5
Very superior	0	0	0	0	3	3

TABLE 9.3 Standard and Potential Profiles of Children Aged 12 to 16

Range of Intellectual Functioning	*Score*					
	Full Scale		*Verbal*		*Performance*	
	Standard	*Potential*	*Standard*	*Potential*	*Standard*	*Potential*
Mentally retarded	5	0	7	0	5	0
Deficient	5	3	21	1	5	1
Borderline	34	3	32	19	16	5
Low average	31	28	29	29	25	8
Average	22	50	11	48	39	49
Above average	7	11	4	7	8	17
Superior	0	9	0	0	4	11
Very superior	0	0	0	0	2	13

Analysis of the data revealed that more than 80% of the children showed evidence of cognitions related to memory, reasoning, and knowledge. They did grocery shopping, performed complex housekeeping tasks, repaired electrical appliances and automotive vehicles, and engaged in conversations with their siblings, peers, and adults that revealed a rich source of information and vocabulary. In short, on tasks of everyday cognitions, these children excelled—a finding for which there is substantial supportive evidence in the literature of both experimental and cultural psychology and anthropology (Cole, Gay, Glick, & Sharp, 1971; Lave, 1977; Rogoff and Waddell, 1982; Saxe, 1988; Serpell, 1979).

Other Intelligences Data

As described in Chapter 6, from our early informal conversations with parents and observations of children we speculated that children were far more cognitively proficient than what the IQ measure revealed. We surmised that if children did indeed possess other intellectual competencies, then elicitation of them may provide two useful functions: (a) boost the children's self-image as cognitively competent beings and (b) knowledge of their own competencies may serve as a motivational carrot to encourage them to engage in tasks in which their IQ-like competencies appeared to be less than optimally developed.

The Other Intelligences Inventory was administered to the parents, teachers, and the children themselves, who comprised the second sample (Table 9.5). The analysis of data revealed consistency of the reports from all three groups. Again, more than 80% of the children were reported to possess intermediate and advanced intellectual competencies in music, dramatization, dance, and other bodily-kinesthetic domains. Table 9.4 and 9.5 present a description of the background characteristics of the sample and a breakdown of the other intelligences findings.

Other Support for the Bioecological Assessment System

Psychologists in Training

Students in graduate programs in clinical and school psychology have used the bioecological assessment system as part of their pre-

Table 9.4 Background Characteristics of Children

Cultural Background						*Language*					
Anglo	*African American*	*Spanish*	*French*	*English*	*Other*	*Spanish/ French*	*English*	*Other*	*Spanish/ French*	*English*	*Other*
23	3	14	0	3	7	11	24	0	4	47	0

Table 9.5 Performance on Other Intelligences Scale

Bodily Kinesthetic ($n = 37$)			*Musical* ($n = 10$)			*Spatial* ($n = 4$)		
Beginner	*Intermediate*	*Advanced*	*Beginner*	*Intermediate*	*Advanced*	*Beginner*	*Intermediate*	*Advanced*
4	18	15	2	3	5	2	2	0

paratory fieldwork experience in intellectual assessment. During initial training, some students had reservations about the use of the psychometric potential measure, admitting that they felt that they had "sinned" when they stepped away from standard procedures for the IQ administrations. Once they were fully trained with the entire system and saw the diagnostic and prescriptive power of the approach, however, they were more comfortable with the concept of "stepping away" from formal assessment of intelligence.

End of semester evaluations revealed that students felt more confident in their ability to conduct more culturally sensitive assessments and to appreciate the diagnostic and prescriptive utility of such an approach.

Psychologist-Practitioners

During the 1995 and 1996 academic year, we conducted a series of workshops on the Biocultural Assessment System for the New York City Board of Education for practitioner-psychologists at their regional offices in the five boroughs. An evaluation questionnaire was administered to each group following completion of the workshop. From a sample of 175 participants, more that 80% indicated the following:

- Their knowledge was broadened regarding the cognitive capabilities of students, particularly those with linguistically and ethnically diverse backgrounds.
- The assessment system had merit for writing the psychological report in a manner that included more diagnostic information about the child's cognitive functioning.
- They had a greater appreciation for qualitative information to complement the psychometric description of the child's cognitive functioning.

Many of them, however, expressed reservations that the assessment may have limited usefulness unless changes were made in policy guidelines for administration of standardized tests of intelligence.

Conclusion

The evidence to date is promising with respect to the development of the four-tier approach to assessing intelligence. The psychometric potential procedures have yielded consistent evidence of the Vygotskian zone of proximal development—that is, the difference in performance when the psychometric and psychometric potential procedures are used. The qualitative information acquired through the ecological taxonomy provided further confirmation for our notion that cognitions are shaped by experiences in particular cultural niches. It is not that some children do and others do not possess the cognitions tapped on traditional tests of intelligence, as IQ scores on the IQ measure suggest. Rather, to make this determination far greater diagnostic probing in multiple ecologies in which the child functions is required than is currently allowed in any standardized intelligence testing context. We were also encouraged by the findings from the Other Intelligence Inventory that suggest that, at least for some children, the experiences and the ecologies that sustain them seemed to have nurtured some cognitions and not others. As we indicate in Chapter 11, we are continuing to refine our procedures in an effort to generate stronger validity and standardizability for the assessment system. Currently, we are exploring the generalizability of our procedures with other commonly used psychometric measures of intelligence. We hope that in time we will have the kinds of empirical data that are both necessary and essential for the emerging biocultural perspective of intelligence that we espouse.

PART III

Training and Policy Implications of the Biocultural Assessment System

10

Training of Mental Health Workers, Educators, and Parents to Enhance the Intellectual Functioning of Children

Can Intelligence Be Taught?

Many avid supporters of IQ tests contend that IQ cannot be boosted (Herrnstein & Murray, 1994; Jensen, 1969). Over the years, however, a number of programs have mushroomed that are designed to increase intellectual skills. These programs are based on the assumption that intelligence is dynamic and mutable and as such can be enhanced through intervention strategies. The more notable programs in which significant gains were seen include Abel (1973),

Adams (1989), Armour-Thomas (1992), Budoff (1987b), Feuerstein's Instrumental Enrichment (1980), Herrnstein, Nickerson, de Sanchez, and Swets (1986), Kornhaber, Krechevsky, and Gardner (1990), Moyer (1986), Scarr and Ricciuti (1991), Sternberg (1985a, 1985b, 1986), Sternberg and Davidson (1989), Thompson and Hixson (1984), Whimbey (1975), and Whimbey and Lochhead (1982). A rather interesting conflict was noted in the works of Richard Herrnstein, coauthor of *The Bell Curve,* which generated much discussion and controversy. In 1986, Herrnstein et al. found that, after working with 400 Venezuelan seventh graders, "Cognitive skills can be enhanced by direct instruction" and that all of the researchers "came away with the strengthened belief in the possibility of teaching intellectual competence more directly than conventional school subjects do" (p. 1289). In his book on the Bell Curve (Herrnstein & Murray, 1994), however, no mention was made to his earlier findings with this cross-cultural population. In any event, we undoubtedly believe that a child's intelligence can be boosted through teaching and coaching, which can be accomplished through several significant groups of people—mental health workers, parents, and teachers.

Training for Psychologists

Assessment

In addition to the issues raised in Chapters 6, 7, and 8, psychologists in graduate training must receive two semesters of training in assessment; the first must provide exposure to psychometric tests such as the Wechsler scales and any other tests that the institution deems necessary. Thus, a prerequisite for training in the biocultural assessment model is a course on psychometric intellectual assessment. It is in the second semester of training in the assessment of intelligence that the biocultural assessment system should be introduced. It is important to remember that only after one has been taught the psychometric properties can one fully appreciate their limitations and when to "step away" from standardized procedures in assessment.

On the basis of experiences of the authors, training in the biocultural assessment system can take place in one semester. The following is an example of what was found to be most effective:

Lecture 1: This lecture provides a review of psychometric testing and an overview of the biocultural assessment system.

Lectures 2, 3, 4, and 5: These lectures provide an exploration of the second tier—psychometric potential assessment. Each component within this tier should be examined in one lecture. Time should be allotted for role play and case samples.

Lectures 6, 7, 8, and 9: These lectures provide an exploration of the third tier—ecological assessment. Each component within this tier should be examined in one lecture. Time should be allotted for role play and case samples.

Lectures 10 and 11: These lectures provide an exploration of the fourth tier—other intelligences. A lecture should be allotted for musical intelligence and for bodily kinesthetic intelligence.

Lectures 12, 13, and 14: These lectures discuss report writing, implications for classroom intervention, and so on. Case reports should be discussed as they apply to this model.

In addition, the curriculum should take on a more interdisciplinary approach, utilizing the contributions from related fields such as social work, psychiatry, and anthropology. Intradiscipline by way of exposure to cross-cultural issues in clinical, counseling, social, developmental, and educational psychology can be quite beneficial in cross-cultural training. Moreover, areas such as psycholinguistics, bilingual and multicultural education, cross-cultural theory, and cross-cultural assessment are all necessary requisites in developing this interdisciplinary competence. Ethical and legal issues in multicultural assessment, consultation, supervision, research, and so on should be infused in each course. Ridley (1985) suggested that every effort should be made to ferret out the principles that are universal in nature so that a basis for determining where cultural variability begins and cultural generalization ends would be established. Exposure to various cultural groups should afford students the opportunity to be part of a viable programmatic experience.

Further Training for Psychologists and Other Mental Health Workers

Treatment Intervention: Cognitive and Behavioral Therapy

Whimbey (1975) proposed a cognitive therapy approach to training intelligence whereby children could be trained in concept formation, classification, categorization, generalization, a graded series of block designs, analytic reasoning, sequential analysis, and sequential deduction. Tasks such as "if-then" reasoning in everyday situations could be explored. Teaching children to be reflective before responding, to use a sort of Socratic dialogue to encourage reflective thinking, and to foresee consequences is critical to the training of intelligence. Cause and effect reasoning stimulated through questions, such as "Why do we need to stay indoors if it is raining?" helped build sequential thinking skills.

Gains in the IQ score of children from the Milwaukee Project in Wisconsin were reported when their intelligences were trained via a three-part process—language development and expression, reading, and mathematics and problem solving. In general, these training techniques increased the children's IQ score between 9 and 15 points. Through lengthy Socratic discussions in an 8-week training program, Gopaul-McNicol and Armour-Thomas (1997b) offer suggestions to mental health workers to further advance the cognitive skills of children through exposing children to intelligence tests items. Whimbey (1975) referred to this as "teaching test-taking ability" and "teaching intelligence since the capacity to analyze problems in this way is exactly what intelligence is" (p. 61).

Training for Parents

Via the Portage Project, Shearer and Loftin (1984) propose a guide for teacher and community leaders on how to assist parents to teach their children to enhance their potential at home. Structured and informal activities for the parent and the child that are generalized to the community at large are offered. Sternberg (1986) offered several strategies for enhancing the memory of children, including categorical clusters whereby an individual is taught to group things by categories instead of trying to memorize in an unordered fashion. Interactive imagery is another technique to aid in memorizing objects or events. If the items cannot fit into a convenient category, Sternberg recommends generating the unrelated words in interactive images such as by the method of loci. Of course, remembering objects by forming acronyms by noting the first letter of each word and making an acronym can increase memory.

Whitehurst et al. (1991) found that severe language problems in children can be ameliorated with a home-based intervention that uses parents as therapists. Parents were given seven standard assignments on a biweekly basis that lasted about 30 minutes each. Role play and other behavioral interventions aided the children in increasing their expressive vocabulary, and this was generalized to other situations and maintained over time.

Rueda and Martinez (1992) proposed a "fiesta educativa" program whereby parents play an active role through community programs to address the needs of their learning disabled youngsters. Essentially, many Latino families worked in tandem to oversee the assessment process, the remedial services, and the overall mental health services. Educating the parents on their children's educational rights was a critical component of this program. Strom, Johnson, Strom, and Strom (1992a, 1992b) found that schools can better serve communities when opportunities for growth are provided to both parents and children. The main point is that Latino parents can help enhance their children's intellectual skills by encouraging their children to ask more questions and to experiment with problem solving in a more independent fashion. Allowing their children the freedom to engage in fantasy and play was also an important characteristic for enhancing intelligence. In general, it was found that children's divergent and convergent thinking, memory, and creative problem solving were increased by teaching these skills through a 4-week (each session lasting 2 hours) parent curriculum, which was as follows:

1. The first session focused on the folly of defining giftedness via a single criterion. Then a more comprehensive perspective (Gardner's multiple intelligences) was presented. Before the end of this session, par-

ents were taught how to identify their children's other intelligences, skills, and gifts.

2. The second session dealt with the kinds of activities that teachers can use in the classroom to enhance critical thinking. An individualized instructional plan for each child was shared with parents. Adequate time was allotted for questions and answers.
3. Session three allowed parents to identify their own strengths and to evaluate their ability to be tolerant of persistent and inopportune questions raised by their children and their ability to be supportive of their children engaging in conversation with adults.
4. Session four gave specific handouts to parents on how to continue enhancing their children's intellectual potential. Guidelines for follow-up sessions were given so parents could continue to be supportive to each other after the group had terminated through a type of steering committee.

We endorse programs such as that outlined previously. We propose, however, a longer training period for parents—an additional 4 more weeks during which parents are taught to enhance their children's intellectual skills by exposing them to tasks commensurate to the type of tasks found on IQ tests and then generalizing these skills to the classroom. Cultural transmission from the home to the school is essential for optimal functioning. Therefore, if the results reveal that a child has a particular concept in one way, but the school ecology needs it to be reflected in another, then it is incumbent on the parent and the school official to train that child to master the skill in the way the school desires. For instance, when a child can put a fan together and is unable to put pieces of a puzzle in a unified whole—tasks that are conceptually quite similar—then such a child can be directly taught through mediated learning experiences (Feuerstein, 1980; Lidz, 1991) how to transfer this knowledge from one context to another. Thus, children can be exposed to puzzles, blocks, and sequential types of tasks such as storytelling via pictures and games that have different and similar features to help nurture abstract thinking. Parents should be encouraged to teach their children to remember in a rote manner their timetables (as is done in the British educational system) so as to develop the ability to do computations mentally. Gopaul-McNicol and Armour-Thomas (1996) offer a step by step practical guide to parents, teachers, and community members on how best to nurture and enhance the intellectual abilities of children. Moreover, the *Guidelines for Providers to the Culturally Diverse* (American Psychological Association, 1993) offers culturally relevant suggestions for practice. More community visits during which contact is established with families, community leaders, and church representatives are critical in understanding the learning styles of children and in using these systems as support to aid in the best assessment practices of children. In general, the research supports that children can be trained to think creatively and enhance their intellectual potential if the appropriate intervention is put in place.

Training for Teachers

Teachers' implicit theories of children's intelligence help to shape the manner in which they respond to them in the classroom (Murrone & Gynther, 1991). We found that teachers were more demanding of children with above-average IQ scores as measured by standardized tests of intelligence. As such, teacher attitudes and perceptions of intelligence tests scores need to be changed through a reeducation process. Maker (1992, p. 32) emphasized that "Not only do standardized tests not predict success in nonacademic settings, but they also are poor predictors of success in school." Maker (1992) also found that the intelligences of children can be enhanced by teaching them Tangram activities (logical mathematical reasoning) in an enrichment program.

Riley, Morocco, Gordon, and Howard (1993) examined what it takes for complex ideas to become rooted in the daily instruction of teachers. Therefore, the authors explored how teachers could design their curriculum to include the needs and strengths of all students. They recommended analog experiences (writing and reading in different genre, conferencing, and role playing) to activate the children's higher cognitive abilities. They also recommended posing questions to children in a directive manner. Thus, children were always expected to develop their responses in a more elaborative type of response.

Armstrong (1994) expanded Gardner's (1993) multiple intelligences in the classroom and in so doing aided teachers in enhancing the varied skills of all children. The concern of proponents of the multiple intelligences theory is that "traditionally, schools have focused on students' analytic, mathematic, and linguistic intelligence which comprise a general intelligence as measured by the IQ test" (Murray, 1996, p. 46). Contrary to this psychometric school of thought, the other intelligences of children are being nurtured as a form of recognizing, respecting, nurturing, and enhancing the holistic intellectual potential of all children.

Authentic Teaching, Learning, and Assessment for All Students (ATLAS) is a comprehensive reform program that combines the work of four organizations: the Coalition of Essential Schools, the School Development Program, the Educational Development Center, and the Development Group of Project Zero. ATLAS emphasizes all the initiatives of these organizations—personalized learning environment, home-school collaboration, an active hands-on type of learning, and ongoing assessment through a curriculum-based approach that responds to the students' strengths.

Lee Katz (1991) spoke of the home-school connection. She spoke of scripts that we all acquire through our experiences and through various contexts. Many children are socialized in the home to a particular script and, when they enter the school setting, the script is different. Therefore, a child who was taught to be emotionally expressive in his or her adult-child interactions at home and then comes to the school, in which the interaction is emotionally cool, must be taught a new script. This is commonly the case with many African American children (Allen & Boykin, 1992), who were found to have more emotionally expressive experiences in their homes. The school psychologists, the teachers, and the special education prevention specialists can assist such children in understanding when one script is preferred over the other. In other words, the idea should not be to inform the child that his or her script is inferior but rather that in the school setting, he or she must recognize when to use which script. The child has to be taught the various scripts that he or she can use. This is analogous to a bilingual child who learns that in the classroom he or she speaks English but can engage his or her peers socially in his or her native language. If we are going to improve learning for all children, the teachers, parents, and significant others must work in a collaborative manner to bring the scripts from the home, the community, and the school closer together.

Adams (1989) offers a thinking skills curricula that first include ecologically valid materials, such as real-world experiences, followed by more abstract materials typically found on psychometric tests. *The Odyssey: A Curriculum for Thinking* (Adams, 1989) focused on the foundations of reasoning, understanding language, verbal reasoning, problem solving, decision making, and inventive thinking in a seven-part creative thinking program.

Armour-Thomas and Allen (1993) developed a cognitive training-intervention program based on Sternberg's triarchic theory of intelligence. The purpose of the program was to help teachers understand the nature of cognitions, in this instance, (a) Sternberg's (1986) metacomponents, performance components and knowledge-acquisition components; (b) the function of these cognitive processes in student's learning; and (c) the importance of explicitness in the use of thinking processes in three major areas of teachers' work—instructional objectives, teacher-student interactions during instruction, and assessment.

Evaluation of the program revealed certain characteristics of teachers classified as high users of process: (a) There was a consistency in their high use of process in all three stages of teaching—the objectives for their students were process focused; (b) the interactions with students during instruction were also process-focused. The kinds of questions they asked and the quality of the feedback given to students demonstrated that not only did they model the process but also they encouraged student awareness and use of these processes; and (c) the emphasis on process was also apparent in the way they designed their assessment procedures—variation in the format, variation in the level of complexity of the tasks, and content equivalent to what students had learned in class.

A multidisciplinary approach to enhancing the performance of children has been touted as the new model of the millennium. Haynes and Comer (1993) recommend a theme concept approach to address the needs of children. At the Yale University Child Study Center, this team of professionals works closely with the home in a collaborative manner. This School Development Program is at the foundation of a holistic development perspective developed by James Comer, now known as the Comer Process for Reforming Education (Comer, Haynes, Joyner, & Ben-Avie, 1996). This model looks to the mental health team, the central organizing body in the school, to involve parents and teachers alike in a decision-making capacity to address the sociocultural needs of the child. This approach is one of collaboration rather than autocracy. Parents are selected by their fellow parents to represent their views on school planning. This indeed bridges the gap between the home and the school. We endorse all the previous initiatives and in particular we emphasize the importance of including people from the community in effecting these changes. We are increasingly mindful that the classroom is quite different from the context of the research lab, which is decontextualized and free from the ongoing activity of the typical classroom. Also, although teachers are trying to include a more dynamic approach to tutelage, they are also dealing with the increase in student diversity that is making teaching the most challenging vocation of our time. Enabling a high level of proficiency in cognitive competence in students will require much more than training in the use of process-based pedagogical strategies in planning, instruction, and assessment—the major areas of the teacher's work. In addition, teacher training programs need to provide experiences for teachers to

appreciate the cognitive strengths that children bring to the classroom;

think of cognitive weaknesses as experience specific and not as general person-specific deficits;

explore ways by which children's everyday cognitions could be applied to school tasks;

engage teachers in self-reflective practices in which they confront their beliefs about children whose cultural socialization may be different from theirs.

These experiences are likely to be rewarding to the extent that training programs

forge more meaningful collaborations with the home and community so as to more fully appreciate the other cultural niches in children's lives;

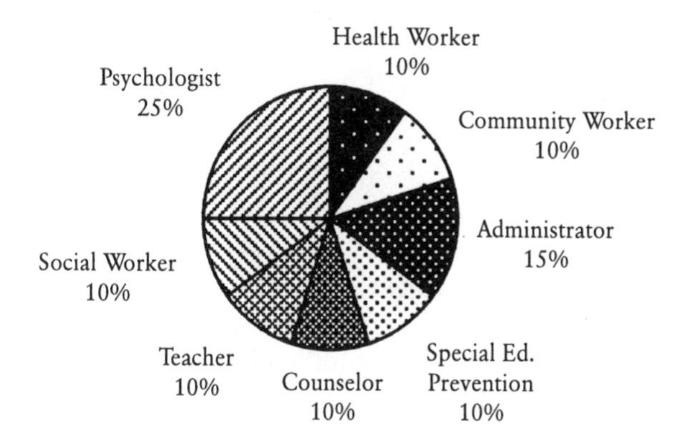

Figure 10.1. A Multisystem Interdisciplinary Model

ensure that teachers are supported with extra resources in the classroom so that they may more effectively apply information gathered from the biocultural assessment system.

As such, we recognize the need to share the responsibility in the schools. We envision the psychologist as playing a prominent role in linking all the disciplines in a multidisciplinary team approach as outlined in the following section.

Psychologists' New Role as Consultant to Parents, Teachers, and the Special Education Prevention Specialists

Currently, in the school and clinical systems there are many disciplines that function as multidisciplinary teams in that each discipline works almost as a separate unit and comes together primarily at Committee on Special Education meetings mainly to decide on placement for the child. We hope that the various disciplines can begin to see the need to be more interdisciplinary than multidisciplinary (Figure 10.1).

Our vision is that administrators such as psychologists can serve as leaders in the school and community systems to bring these various disciplines together and we hope that this will enhance the quality of education and, in particular, bring about a reduction in special education placement.

Thus, with more interdisciplinary training, psychologists can team up with health workers so that school officials will understand how health issues impede or enhance the child's functioning.

Likewise, psychologists can work with all school administrators to see how their assessment findings can be used in a prescriptive manner to assist the teachers in the classroom. At this juncture, special education teachers will intervene as special education prevention specialists. Therefore, with the help of these three subdisciplines, children will learn to use all their strengths to work on their weaknesses. This should result in a reduction of special education placement.

Moreover, psychologists can work with the social workers and community resource people to develop a pool of resources outside of the school settings to best empower the child and his or her family.

If the family is involved in any sort of treatment by psychotherapists, psychologists, psychiatrists, and so on, the psychologists can serve as a liaison and can aid school personnel in understanding how the treatment program is enhancing or impeding the child's progress.

Finally, psychologists can work with guidance counselors and serve in consulting capacity—a sort of mediator between the school and vocational organizations outside of the school setting. This is to ensure that there is a connection between completion of high school and job or advanced educational opportunities.

For school psychology to survive with dignity, a multimodal, multisystems approach is needed to address the needs of people from varying ethnic and cultural backgrounds.

If such an interdisciplinary program is initiated, we envision this type of program becoming the model training program for the 21st century.

Example of an Interdisciplinary Training Program at the University Level

Johnson (1982, 1987, 1990) developed a two-part course that included theory, current research, and a laboratory experiential-type section.

Practica in all areas of their training should be available to ensure that all students receive hands-on training in assessing children, especially culturally and linguistically diverse children. To accomplish an innovative interdisciplinary training program, the following step-by-step guidelines are suggested:

1. A written academic policy emphasizing a clear statement of purpose and commitment to an interdisciplinary cultural diversity type of training as well as the consequences to the program if these policies are violated should be provided. Included in this statement must be definite quantifiable, tangible program objectives that must be achieved during a particular time frame.
2. Cultural and ethnic content should be infused in each course and not taught as a single course. Thus, when teaching psychological assessment, students should first be exposed to adherence of standardized procedures, and then they should be taught how to assess the cognitive potential of children via techniques such as Item Equivalency Assessment, Other Intelligences Assessment, Test-Teach-Retest Assessment, Ecological Assessment, and Suspending Time.
3. There should be a more aggressive recruitment of faculty members and students of various cultural backgrounds. Working with an ethnically diverse student and faculty body adds enrichment to the program because one can view issues from various perspectives.
4. Faculty members should be encouraged to update their interdisciplinary and cross-cultural expertise by attending continuing education courses, seminars, and so on. The university should give a reduction in faculty members' teaching load for 1 year to allow for this type of training.
5. A consultant or a full-time faculty member with interdisciplinary expertise should be available to consult with all faculty members to assist them in redesigning their curriculum to reflect a culturally diverse interdisciplinary content.
6. Funds should be set aside for a few research students to be assigned strictly to this interdisciplinary and cross-cultural thrust—building a resource file, assisting in student and faculty recruitment, linking with community people to recruit more ethnic minority practicum supervisors, coordinating experts of different cultural backgrounds to speak at colloquia, and so on.

7. For the first 2 years, an ongoing review of the program to ensure that the goals are being met should be done on a monthly basis at regular staff meetings. After 2 years of smooth functioning, it should be done on a quarterly basis, and after 5 years, an annual basis should suffice. Every faculty member should be required to sit in on these meetings.
8. Without all of the above in place and without financial support to fund all these innovative efforts, failure is most likely to occur. Because the ultimate goal of innovation should be institutionalization, hard-line financial endowments are needed (Ridley, 1985).

A biocultural interdisciplinary curriculum should therefore be multifaceted, consisting of a combination of assessment, review of the ethnic literature, personal involvement, and the development of a small classroom group project (Parker, Valley, & Geary, 1986). This approach utilizes the cognitive, affective, and behavioral domains. The students should first be assessed on their knowledge, attitudes, and perceptions of interdisciplinary and cross-cultural experiences as well as their comfort level in interacting with others from different ethnic and racial groups. This assessment process serves as a guide to the professor for future training.

Part 2 of the course involves the readings and discussions about the variability within majority and minority groups so that no one ethnic group will be stereotyped. Part 3 is very action oriented in that it involves behavioral activities geared to helping students increase biocultural assessment skills and cultural knowledge, sensitivity, and effectiveness. Students initially observe from a distance videotapes and so on and gradually move to participate directly. In the final stage, the students are expected to work on a small group activity in class. Such a project allows the students to become aware of their own stereotypical values and assumptions about intelligence and about other cultural and ethnic groups.

A true commitment to interdisciplinary and multicultural training requires at a minimum the implementation of all the previously discussed guidelines. If this program has only the bare skeleton of a commitment, then it creates no more than false generosity (Freire, 1970), dishonesty, and continued disrespect. The goal should be to produce competent psychologists capable of working with children from any linguistic, cultural, and ethnic background.

11

Implications for Future Research and Policy

It may be recalled that the biocultural assessment system was developed in response to our concern that standardized tests of intelligence provided an incomplete appraisal of children's cognitive functioning. Guided by the assumptions of our emerging biocultural theory, a number of cognitive enhancement procedures were used in conjunction with the traditional IQ measure—the Wechsler Intelligence Scale for Children-III (WISC-III)—to identify cognitive strengths and weaknesses of children. We are encouraged with the results, which demonstrate that improvements on intellectual tasks are to be expected when children are

allowed opportunity to contextualize words in sentences;

given time, paper, and pencil to solve verbal problems involving memory and quantitative reasoning;

given time and opportunity to understand and solve spatial problems involving memory, understanding, and reasoning.

In addition, we have observed that many children who perform poorly on standardized IQ tasks are able to demonstrate comparable skills in their everyday lives. Finally, we found that many children, including those with below-average performance on the IQ measure, possess other intellectual competencies at varying degrees of proficiency.

Consideration of these results from a biocultural perspective would suggest that a standardized measure of intelligence is insufficiently sensitive to discriminate which cognitions are well formed from those that are still in an embryonic stage of development. Furthermore, such a measure tells us nothing about the prior experiences of children or the ecologies in which they grow and function and how these culturally specific experiences may have influenced the performance observed at the time of assessment. In contrast, the results from the other procedures, psychometric potential and other intelligences, were far more diagnostically useful in probing for both emerging cognitions and those that are already well formed but may be in need of further development. The observation that children who seem "unintelligent" in the testing environment but are cognitively adept in their homes and communities provided additional confirmation of our conception of intelligent behavior as culturally dependent cognitions.

These findings present some interesting challenges for researchers (Table 11.1) regarding the types of questions they may wish to consider regarding intelligence and its assessment as well as the kinds of research design and methodology that such questions would necessitate. We think, however, that the findings pose a more pressing challenge to those who set guidelines and standards for intelligence testing, particularly as these relate to nondiscriminatory practices for children from linguistically and ethnically diverse backgrounds (Table 11.2). This chapter gives an overview of some of these challenges.

Research Implications

IQ Studies

Through administration of the IQ measure, we were able to discern some diagnostic information regarding strengths and weak-

TABLE 11.1 Ideas for Future Research

Develop new theories to examine the relationship between cognitive processes and ecology.
Examine the validity of the current assessment measures that assess linguistically and culturally different children.
Explore the use of the biocultural approach to cognitive assessment with all children.
Train mental health workers to more accurately assess all children.
Design studies in ways that allow the interpretation of the differences found to be in line with the measures used. For example, the Other Intelligences Inventory was developed to address the concept of multiple intelligences.
More longitudinal studies are needed to address the cultural deprivation and biological inferiority positions that exist in traditional research.
Develop research that examines diagnosis as informing prescription not describing deficits—a more enhancement model.

TABLE 11.2 Main Policy Issues for the New Millennium

We need a clear policy regarding the principles and standards of intelligence testing beyond the standardized testing instruments. The challenge is to remove the paradigm from reliance on prediction to understanding prescription.
We need clarity of policy on the federal level. Currently, it is left up to the individual districts and states. Therefore, the American Psychological Association and National Association of School Psychologists should support such a policy because many decisions are made on children's lives that center around the IQ tests.
We need policies that provide clear standards for desirable competencies among practitioners, both psychologists-in-training and practicing psychologists and clinicians.
We need a clear policy regarding the more dynamic role that psychologists should play in guiding the teacher, parent, and other school staff in best serving all students.
We need to have a policy regarding the role of the special education prevention specialists, who can serve as mediators between the classroom and special education.

nesses in cognitive functioning as indicated in Chapter 9. It is quite likely, however, that far more educationally meaningful information may be derived from an IQ test than we were able to extract. Gordon (1977, 1995) offers the following step-by-step procedure for the analysis of test scores that may yield other diagnostically useful information:

1. Identify, through logical analysis, the dimensional or functional demands of selected standardized tests.
2. Determine the rationale utilized in the development of each of several tests to identify the conceptual categories for which items were written and in which item response consistencies might cluster empirically.
3. Determine the learning task demands represented by the items of selected tests and classify those demands into functional categories.
4. Appraise the extent to which selected tests provide adequate coverage of the typical learning task demands found in educational settings.
5. Utilize the categories produced by any or all of the previous strategies in the metric and nonmetric factorial analysis of test data to uncover empirical dimensions of test responses.

The intent of the first four tasks is to unbundle the cognitive competencies embedded in test items. The fifth step involves the actual analysis of performance data to reveal their factorial demand structure and to ascertain the extent to which they yield empirical evidence for those item clusters or require a reconceptualization of response processes. Assuming that there is congruence between the fifth task and the other four tasks, the resulting data may reveal diagnostic patterns that then become the basis for prescriptive instruction.

Other Intelligence Studies

We were able to identify consistently other intellectual competencies that children possessed in varying degrees of proficiency. If, as a society, we value these intelligences as adult end states, then it is important for research to seek a more informed understanding of the content and context of their development. For example, if musicians, mathematician, dancers, computer scientists, and aerospace engineers are valued by a society, then opportunities should be made available for the development of intellectual competencies related to the knowledge and skills related to these professions. The work of Gardner and Hatch (1989) in the Arts PROPEL and Project Spectrum demonstrates the kinds of experiences in multiple domains of knowledge that are likely to nurture the growth and development of spatial, linguistic, musical, and other intellectual competencies. Apprenticeships or mentoring relationships are other types of learning experiences in particular domains of knowledge (e.g. architecture, visual arts, and music) whereby emerging intelligences can be nurtured and developed. Schools, community organizations, and other institutions of a society, such as art, music, math, and science museums or galleries, provide a rich variation of tools, materials, and culturally meaningful information from which children can acquire and apply different forms of knowledge. Longitudinal studies could be designed that begin with the collection of baseline data on children's emerging intellectual competencies in cultural niches wherein certain types of experiences are valued and are made available by the community. Subsequently, at different periods in time, these person-environmental interactions could be assessed to better understand the nature and quality of change and continuity of the nascent intellectual competencies observed earlier.

Cognitive Enhancement Studies

We used a variety of strategies to elicit cognitive potential masked by the standardized IQ measure. To the extent that these competencies represent meaningful adult end states in a society—and they do in the U.S. culture—then every effort should be made to foster these malleable cognitions in the teaching and learning experiences both within and outside the classroom. We concur with the theoretical and empirical work of a number of researchers (e.g., Feuerstein, 1990; Feuerstein, Rand, & Hoffman, 1979; Gardner, 1983; Lave & Wenger, 1991; Sternberg, 1986; Vygotsky, 1978; Whimbey & Lockhead, 1982) regarding the trainability of intellectual skills. We think, however, that the design of cognitive enhancement studies should ensure that essential and sufficient opportunities are provided for transfer of cognitive competencies to other tasks and settings. The following suggestions are made for critical person-task-context conditions if intervention effects are to be expected:

1. There must be opportunity to acquire and use knowledge and cognitive skills in the preferred symbol system of the cultural group.
2. There must be opportunity for the social mediation (direct and indirect) of acquisition of knowledge and cognitive skills within the selected symbolic system of the cultural group.
3. There must be opportunity for independent practice of knowledge and cognitive skills within the symbol system in which the initial knowledge and cognitive skills were acquired.
4. There must be opportunity for social mediation (direct and indirect) in the application of previously acquired knowledge and skills in relatively new and more complex tasks represented through a familiar symbol system.
5. There must be opportunity for independent use of previously acquired knowledge and cognitive skills in relatively new and cognitively more complex tasks represented in a familiar symbol system.
6. There must be opportunity for social mediation (direct and indirect) in the application of previously acquired knowledge and skills in relatively new and progressively more complex cognitive tasks represented in an unfamiliar symbol system.
7. There must be opportunity for independent practice of previously acquired knowledge and cognitive skills in relatively new and progressively more complex cognitive tasks represented in an unfamiliar symbolic system.

These recommendations assume that generalizability of learning from one context to another is contingent on careful manipulation of the person-task-context variables and is consistent, for the most part, with the strategies identified in the cognitive science literature (e.g., Anderson, Reder, & Simon, 1996).

Perhaps, as important, intervention studies should be longitudinal in design to allow sufficient time for reinforcement and generalizability of effects to other contexts. It may well be that Head Start and other intervention studies for improving intelligence showed little enduring results due to the brevity of the treatment. During the course of development of some children, threatening person-environment interactions may far outweigh the sustaining person-environment ones. To offset the negative impact of the former, longer and more enriching interventions are likely to produce more lasting cognitive change. As Horowitz and O'Brien (1989) noted,

> Development is not a disease to be treated. It is a process that needs constant nurturance. There is no reason to expect that an intensive program of early stimulation is an inoculation against all further developmental problems. No one would predict that a child given an adequate amount of vitamin C at 2 years of age will not have vitamin C deficiency at 10 years of age. Currently, according to the most viable models of development that apply to both at-risk and normal children, developmentally, functional stimulation is desirable at every period of development and not only in early years. (p. 444)

Ecology Studies

The prescriptive utility of our findings, particularly as they relate to schooling, was identified in Chapter 9. These interventions, if explored, are unlikely to lead to enduring results unless sustaining conditions and forces are operating within ecologies beyond the school in which children grow and function. More than 30 years ago, Coleman et al. (1966) called attention to relative influence of schooling on the cognitive outcomes of children by stating the following:

> The school brings little influence to bear on a child's achievement that is independent of his background and general social context; and that this very lack of an independent effect means that the inequalities imposed on children by their home, neighborhood, and peer environment are carried along to become the inequalities which control life at the end of school. For equality of educational opportunity through the school to be effective, one must imply a strong effect of schools that is independent of the child's immediate social environment, and that strong independent effect is not present in American schools. (p. 325)

We know from previous research that socialization practices of the peer group (e.g., Steinberg, Dornbusch, & Brown, 1992) and the home (e.g., Allen & Boykin, 1991) both directly and indirectly influence children's behavior in school. This would suggest that researchers seeking to understand the impact of in-school cognitive intervention programs should simultaneously examine the nature and quality of experiences in the primary ecologies in which children function (e.g., the home and the peer group). In other words, the researcher needs to determine whether sustaining or threatening person-environments interactions or both operating outside the school are likely to reinforce or weaken the effects of the in-school intervention program. It is hoped that the use of such a methodology would minimize a common knee-jerk interpretation in terms of biologically constrained cognitive abilities for the nonsignificant or short-lived effects of many cognitive enhancement programs.

Personal Characteristics Studies

Although our sample came from diverse linguistic, ethnic, and racial backgrounds in the United States, further research is needed on the generalizability of the biocultural approach to intelligence testing to societies beyond the culture of the United States. Also, in generating the data we did not control for individual or groups with contrasting characteristics (e.g., race, gender, ethnicity, or socioeconomic status) because that was not the focus of our work as was explained in Chapter 10. We know from previous research, however, that demographic characteristics (Boykin & Toms, 1985; Gaines & Reed, 1995; Gordon, Marin, & Marin, 1991; Ogbu, 1986; Phinney, 1996; Sue, 1991) and response tendencies (Boykin, 1979; Gordon, 1988; Hale-Benson, 1986; Hilliard, 1992; Shade, 1982) play a crucial role in cognitive behavior. Indeed, we suspect that these person characteristics render some individuals resilient or vulnerable toward environmental encounters and stimuli and as such should be carefully considered in any interpretation of differences in observed cognitive performance. Thus, researchers investigating differences in intelligent behavior should describe samples at a sufficient level of detail so as to better understand the relative contributions of person characteristics in the observed differences in performance.

Instrument Refinement and Development

Through the application of our psychometric potential techniques, ecological taxonomy, and other intelligence measures, we are confident in our findings that there is more to intelligence than the traditional IQ test measures. As we gain a more informed understanding and appreciation of cognitions developed through person-environment interactions in multiple cultural niches in which children develop and function, however, we need to ensure that our measures are ecologically valid as well. Currently, we are refining some of the measures in the Other Intelligence Inventory and expanding the ecological taxonomy to include observational procedures of other intelligences.

Policy Implications

Over the years, professional organizations, such as the American Psychological Association and the National Council of Measurement in Education, have reflected concerns regarding discriminatory practices in intelligence testing through their bylaws, ethical principles, and standards. These noble sentiments, however, seemed to have had minimal impact on the construction of standardized tests or their practice. As indicated in Chapter 6, interpretation and use of test results have been particularly inimical to ethnic and linguistic children, particularly those from low-income backgrounds. It is as if test developers and practitioners have been oblivious to the theoretical and empirical research regarding the situatedness of cognition during the past two decades or to the influence of linguistic and cultural diversity in intelligent behavior. Even more disheartening, however, they seem unaware of the devastating consequences of their judgments that place large numbers of children on educational paths that are neither enabling nor worth wanting. Indeed, the literature is replete with inequitable schooling of children placed in low-track classes or in unwarranted special educational programs (e.g., Kozol, 1991; Lipsky & Gartner, 1996; Oakes, 1990; Skrtic, 1991). Perhaps, a list of principles and standards, no matter how well intentioned, is insufficient insurance against discriminatory practice in standardized intellectual testing. What is needed, in our judgment, is greater democratic reciprocity in discussions among practitioners, test developers, and client representatives in the development of principles and standards setting with respect to intellectual assessment practices. Equally important is the need for enforceable principles that are truly reflective of our commitment to equity and cultural pluralism. Though not exhaustive, in the following sections we submit recommendations for consideration.

Principles Development and Standards Setting

Professional organizations in psychology and education, through their principles and standards, have provided ethical guidelines for practitioners. It is not always clear, however, that the terms of conversation as well as the content of deliberations genuinely reflect the views of parents and community stakeholders who have had first-hand evidence of the deleterious effects of standardized testing of intelligence. Also not clear is whether the principles and standards emerged from the kinds of discourse that Moon (1993), Boyd (1996), and Rawls (1993) identify as critical for the commitment to the principle of cultural pluralism. In other words, have participants engaged in discussions of reasonable pluralism about incompatible conceptions of the human good or ideals of excellence or the dilemma of cultural relativism and universal ethical principles?

These and other questions are the kinds of uncomfortable moral challenges that, in our judgment, should form the agenda for policy-makers if the hidden ugliness of multiculturalism is to be truly unmasked. We submit for consideration the following questions:

Is there representation of diverse epistomologies at the discussion table?

Do the terms of conversation allow for genuine and open discussions of culturalism, pluralism, and universal ethical principles?

Are the principles and standards essentially and sufficiently accommodating of multiple cultural perspectives?

Are the principles and standards supportive of multiple expressions of intelligent behavior?

Are the principles and standards supportive of the assessment of cognitive potential?

Are the principles and standards supportive of evidence of intelligence beyond the standardized testing context?

Is there a mechanism for principles and standards revision or modification?

Do the principles and standards provide clear implications for desirable competencies among practitioners, both psychologists-in-training and practicing psychologists and clinicians?

Do these competencies include skill in intellectual assessment within and outside the standardized testing context?

Do these competencies include skill in assessment of cognitive potential?

Do these competencies include skill in report writing that includes both quantitative and qualitative evidence of intellectual functioning?

Are there mechanisms for the incorporation of these competencies in accreditation criteria?

Principles Enforcement Strategies

Most psychologist-in-training programs have vision and goal statements that describe a commitment to the principles of cultural diversity. In addition, such programs can identify courses with cultural diversity topics and reading lists dealing with the topic. To ensure that the principles and standards as enunciated by the professional organizations are operationalized in practice, however, satisfactory answers should be sought for the following questions by accreditation teams when they visit training programs:

1. Do the end-of-training competencies include concrete evidence of understandings of diverse epistomologies or worldviews?

   appreciation and genuine respect for cultural pluralism?

   culturally sensitive strategies used in assessment?

   skill in gathering data on intellectual functioning within and beyond the testing context?

   skill in analyzing quantitative and qualitative data from assessment?

   skill in writing psychological reports with diagnostic and prescriptive utility?

2. Do courses on intelligence and its assessment explicitly indicate theoretical assumptions about intelligence?

   limitations of standardized tests of intelligence in terms of culturally compatible fallacies?

3. Do supervisors have demonstrable competencies commensurate with those of the end-of-training competencies of the psychologist-in-training?

Conclusion

The biocultural assessment system has greater diagnostic and prescriptive utility than any single standardized measure of intelligence. The data are highly relevant and useful for instructional purposes both within and outside the classroom. It has yielded information from which testable research hypotheses could be made regarding the cultural dependency of human cognition in terms of both its development and its teachability. The early promise of these findings, however, is likely to be ignored or dismissed as have those of other researchers before us unless all those connected to the intelligence testing movement do more than pay lip service to the country's commitment to equality and cultural pluralism. As the new millennium approaches, the question is clear: Are we, the community of custodians of the nation's children, ready and able to provide equitable educational opportunities through culturally responsive assessments or will we, like other generations of custodians before us, shirk our responsibility to do right by those entrusted in our care for service? It may be helpful to remember that the calling of stewardship requires neither scientific evidence nor protective policies as a prerequisite to action but only the moral impulse to do right, especially by those most vulnerable and placed at risk in our society. We are optimistic that the nobler and gentler side of this generation's community of custodians will emerge and that greater efforts will be made to provide education that is both enabling and worth wanting for more of the nation's children.

References

Abel, T. M. (1973). *Psychological testing in cultural contexts.* New Haven, CT: College & University Press.

Adams, M. J. (1989). Thinking skills curricula: Their promise and progress. *Educational Psychologist, 24*(1), 25-75.

Allen, B., & Boykin, A. W. (1991). The influence of contextual factors on black and white children's performance. Effects of movement opportunity and music. *International Journal of Psychology, 26,* 373-387.

Allen, B. A., & Boykin, A. W. (1992). Children and the educational process: Alienating cultural discontinuity through prescriptive pedagogy. *School Psychology Review, 21*(4), 586-596.

American Psychological Association. (1993). Guidelines for providers of psychological services to the ethnic, linguistic and culturally diverse populations. *American Psychologists, 48,* 45-48.

Anastasi, A. (1988). *Psychological testing* (6th ed.). New York: Macmillan.

Anderson, J. R., Reder, L. M., & Simon, H. A. (1996). Situated learning and education. *Educational Researcher, 25*(4), 5-11.

Apple, M. (1979). *Ideology and curriculum.* London: Routledge & Kegan Paul.

Armour-Thomas, E. (1992a). Assessment in the service of thinking and learning for low achieving students. *High School Journal, 75*(2), 99-118.

Armour-Thomas, E. (1992b). Intellectual assessment of children from culturally diverse backgrounds. *School Psychology Review, 21*(4), 552-565.

Armour-Thomas, E., & Allen, B. (1993). The feasibility of an information-processing methodology for the assessment of vocabulary competence. *Journal of Instructional Psychology, 20*(4), 306-313.

Armour-Thomas, E., & Gopaul-McNicol, S. (1997a). The bioecological approach to cognitive assessment. *Cultural Diversity and Mental Health, 3*(2), 131-144.

Armour-Thomas, E., & Gopaul-McNicol, S. (1997b). In search of correlates of learning underlying "learning disability" using a bioecological assessment system. *Journal of Social Distress and the Homeless, 6*(2), 143-159.

Armstrong, T. (1994). *Multiple intelligences in the classroom.* Alexandria, VA: Association for Supervision and Curriculum Development.

Artzt, A. F., & Armour-Thomas, E. (1992). Development of a cognitive-metacognitive framework for protocol analysis of mathematical problem solving in small groups. *Cognition and Instruction, 9,* 137-175.

Asante, M. K. (1988). *Afrocentricity.* Trenton, NJ: Africa World Press.

Banks, W. C., McQuater, V., & Hubbard, J. L. (1979). Toward a reconceptualization of the social-cognitive bases of achievement orientation in blacks. In A. W. Boykin, A. J. Franklin, & J. F. Yates (Eds.), *Research directions of black psychologists* (pp. 294-311). New York: Russell Sage.

Baron, J. (1981). Reflective thinking as a goal of education. *Intelligence, 5,* 291-309.

Baron, J. (1982). Personality and intelligence. In R. J. Sternberg (Ed.), *Handbook of human intelligence.* New York: Cambridge University Press.

Beker, J., & Feuerstein, R. (1990). Conceptual foundations of the modifying environment in group care and treatment settings for children and youth. *Journal of Child and Youth Care, 4*(5), 23-33.

Berry, J. (1976). *Human ecology and cognitive style.* New York: John Wiley.

Betancourt, H., & Lopez, S. R. (1993). The study of culture, ethnicity, and race in American psychology. *American Psychologist, 48*(6), 629-637.

Binet, A., & Simon, T. (1905). Méthodes nouvelles pour le diagnostic du niveau intellectuel des anormaux. [New methods for diagnosing the intellectual level of abnormals]. *Année Psychologique, 11,* 191-336.

Bouchard, T. J., Jr., Lykken, D. T., McGue, M. L., Segal, N. L., & Tellegen, A. (1990). Sources of human psychological differences: The Minnesota study of twins reared apart. *Science, 250,* 223-228.

Boyd, D. (1996). Dominance concealed through diversity: Implications of inadequate perspectives on cultural pluralism. *Harvard Educational Review, 66*(3), 609-630.

Boykin, A. W. (1977). On the role of context in the standardized test performance of minority group children. *Cornell Journal of Social Relations, 12,* 109-124.

Boykin, A. W. (1979). Black psychology and the research process: Keeping the baby but throwing out the bathwater. In A. W. Boykin, A. J. Franklin,

& J. P. Yates (Eds.), *Research directions of black psychologists.* New York: Russell Sage.

Boykin, A. W. (1982). Task variability and the performance of black and white schoolchildren. *Journal of Black Studies, 12,* 469-485.

Boykin, A. W. (1983). The academic performance of Afro-American children. In J. T. Spence (Ed.), *Achievement and achievement motives* (pp. 322-371). San Francisco: Freeman.

Boykin, A. W. (1986). The triple quandary and the schooling of Afro-American children. In U. Neisser (Ed.), *The school achievement of minority children* (pp. 57-92). Hillsdale, NJ: Lawrence Erlbaum.

Boykin, A. W., & Allen, B. A. (1988). Rhythmic movement facilitation of learning in working-class Afro-American children. *Journal of Genetic Psychology, 149,* 335-348.

Boykin, A. W., DeBritto, A., & Davis, L. (1984). *The influence of social process factors and contextual variability on school children's task performance.* Unpublished manuscript, Howard University, Washington, DC.

Boykin, A. W., & Toms, F. (1985). Black child socialization: A conceptual framework. In H. McAdoo & J. McAdoo (Eds.), *Black children: Social, educational, and parental environments* (pp. 32-51). Beverly Hills, CA: Sage.

Bradley, R. H., and Caldwell, B. M. (1984). The relation of infants' home environments to achievement test performance in first grade: A follow-up study. *Child Development, 52,* 708-710.

Bronfenbrenner, U. (1979). *Toward the ecology of human development.* Cambridge, MA: Harvard University Press.

Bronfenbrenner, U. (1993). The ecology of cognitive development: Research models and fugitive findings. In R. H. Wozniak & K. W. Fischer (Eds.), *Development in context: Acting and thinking in specific environments* (The Jean Piaget Symposium Series, pp. 3-44). Hillsdale, NJ: Lawrence Erlbaum.

Bronfenbrenner, U. (1989). Ecological systems theory. In R. Vasta (Ed.), *Annals of Child Development Research, 6,* 185-246.

Brown, A. L. (1978). Knowing when, where, and how to remember: A problem of metacognition. In R. Glaser (Ed.), *Advances in instructional psychology* (Vol. 1, pp. 77-165). Hillsdale, NJ: Lawrence Erlbaum.

Budoff, M. (1987a). The validity of learning potential assessment. In C. S. Lidz (Ed.), *Dynamic assessment: An international approach to evaluating learning potential.* New York: Guilford.

Budoff, M. (1987b). Measures for assessing learning potential. In C. S. Lidz (Ed.), *Dynamic assessment: An interactional approach to evaluating learning potential.* New York: Guilford.

Butcher, J. N. (1982). Cross-cultural research methods in clinical psychology. In P. C. Kendall & J. N. Butcher (Eds.), *Handbook of research methods in clinical psychology* (pp. 273-308). New York: John Wiley.

Carlson, J. S. (1985). The issue of g: Some relevant questions. *The Behavioral and Brain Science, 8*(2), 224-225.

Carraher, T. N., Carraher, D., & Schliemann, A. D. (1985). Mathematics in the streets and in schools. *British Journal of Development Psychology, 3,* 21-29.

Carroll, J. B. (1993). *Human cognitive abilities: A survey of factor-analytic studies.* New York: Cambridge University Press.

Case, R. (1985). *Intellectual development: Birth to adulthood.* Orlando, FL: Academic Press.

Cattell, R. B. (1941). Some theoretical issues in adult intelligence testing. *Psychological Bulletin, 38,* 592.

Cattell, R. B. (1943). The measurement of adult intelligence. *Psychological Bulletin, 40,* 153-193.

Cattell, R. B., & Horn, J. L. (1978). A check on the theory of fluid and crystallized intelligence with description of new subtest designs. *Journal of Educational Measurement, 15,* 139-164.

Ceci, S. J. (1990). *On intelligence . . . more or less: A bioecological treatise on intellectual development.* Englewood Cliffs, NJ: Prentice Hall.

Ceci, S. J., Baker, J., & Bronfenbrenner, U. (1987). *The acquisition of simple and complex algorithms as a function of context.* Unpublished manuscript, Cornell University, Ithaca, NY.

Ceci, S. J., & Bronfenbrenner, U. (1985). Don't forget to take the cupcakes out of the oven: Strategic time-monitoring, prospective memory, and context. *Child Development, 56,* 175-190.

Ceci, S. J., & Cornelius, S. (1989, April 29). *Psychological perspectives on intellectual development.* Paper presented at the biennial meeting of the Society for Research in Child Development, Kansas City, MO.

Ceci, S. J., & Liker, J. (1986a). A day at the races: A study of IQ, expertise, and cognitive complexity. *Journal of Experimental Psychology: General, 115,* 225-266.

Ceci, S. J., & Liker, J. (1986b). Academic and non-academic intelligence: An experimental separation. In R. J. Sternberg & R. K. Wagner (Eds.), *Practical intelligence: Origins of competence in the everyday world.* New York: Cambridge University Press.

Ceci, S. J., & Liker, J. (1988). Stalking the IQ-expertise relationship: When the critics go fishing. *Journal of Experimental Psychology: Human Learning and Memory, 6,* 785-797.

Chi, M. T. H. (1978). Knowledge structures and memory development. In R. S. Siegler (Ed.), *Children's thinking: What develops?* Hillsdale, NJ: Lawrence Erlbaum.

Chi, M. T. H., & Ceci, S. J. (1987). Content knowledge: Its restructuring with memory development. *Advances in Child Development and Behavior, 20,* 91-146.

Cohen, R. (1969). Conceptual styles, culture conflict, and non-verbal tests of intelligence. *American Anthropologist, 71*(5), 828-857.

Cole, M. (1988). Cross-cultural research in the sociohistorical tradition. *Human Development, 31,* 137-152.

Cole, M., Gay, J., Glick, J. A., & Sharp, D. W. (1971). *The cultural context of learning and thinking.* New York: Basic Books.

Cole, M., & Scribner, S. (1977). Cross-cultural studies of memory and cognition. In R. V. Kail & J. W. Hagen (Eds.), *Perspectives on the development of memory and cognition.* Hillsdale, NJ: Lawrence Erlbaum.

Cole, M., Sharp, D. W., & Lave, C. (1976). The cognitive consequences of education. *Urban Review, 9,* 218-233.

Coleman, J. S., Campbell, E. Q., Hobson, C. J., McPartland, J., Mood, J., Winfield, F. D., & Work, R. L. (1966). *Equality of educational opportunity* (No. OE 38001). Washington, DC: U.S. Office of Education.

Comer, J., Haynes, N., Joyner, E., & Ben-Avie, B. (1996). *Rallying the whole village: The Comer process for reforming education.* New York: Columbia University Press, Teachers College.

Cummins, J. (1984). *Bilingualism and special education: Issues in assessment and pedagogy.* San Diego: College Hill.

Cummins, J. (1991). *Bilingualism and special education: Issues in assessment and pedagogy.* San Diego: College Hill Press.

Das, J. P. (1985). Interpretations for a class on minority assessment. *The Behavioral and Brain Science, 8*(2), 228-229.

De Avila, E. (1974, November/December). The testing of minority children—A neo Piagetian approach. *Today's Education,* pp. 72-75.

Detterman, D. K. (1985). Review of Wechsler Intelligence Scale of Children—Revised. In J. V. Mitchell (Ed.), *The ninth mental measurement yearbook* (Vol. 2, pp. 1715-1716). Lincoln, NE: Buros Institute of Mental Measurements.

DeVos, G. A. (1984, April). *Ethnic persistence and role degradation: An illustration from Japan.* Paper presented for the American-Soviet Symposium on Contemporary Ethnic Processes in the U.S.A. and the U.S.S.R., New Orleans, LA.

Dunn, R., & Dunn, K. (1978). *Teaching students through their own behavioral teaching style.* Reston, VA: Prentice Hall.

Eberhardt, J. L., & Randall, J. L. (1997). The essential notion of race. *American Psychological Society, 8*(3), 198-203.

Esquivel, G. (1985). Best practices in the assessment of limited English proficient and bilingual children. In A. Thomas & J. Grimes (Eds.), *Best practices in school psychology I* (pp. 113-123). Washington, DC: National Association of School Psychologist.

Eysenck, H. J. (1982). Introduction. In H. J. Eysenck (Ed.), *A model for intelligence.* Berlin: Springer-Verlag.

Eysenck, H. J. (1986). Inspection time and intelligence: A historical introduction. *Personality and Individual Differences, 7,* 603-607.

Eysenck, H. J. (1988). The biological basis of intelligence. In S. H. Irvine & J. W. Berry (Eds.), *Human abilities in cultural context* (pp. 87-104). New York: Cambridge University Press.

Farnham-Diggory, S. (1970). Cognitive synthesis in Negro and white children. *Monograph of the Society for Research in Child Development, 35*(2), Serial No. 135.

Feuerstein, R. (1979). *The dynamic assessment of retarded performers.* Baltimore, MD: University Park Press.

Feuerstein, R. (1980). *Instrumental enrichment: An intervention program for cognitive modifiability.* Baltimore, MD: University Park Press.

Feuerstein, R. (1990). The theory of structural cognitive modifiability. In B. Z. Presseisen (Ed.), *Learning and thinking styles: Classroom interaction* (pp. 68-134). Washington, DC: National Education Association.

Feuerstein, R., Hoffman, M., Rand, Y., Jensen, M., Morgans, R. J., Tzuriel, D., & Hoffman, D. (1986b). Learning to learn: Mediated learning experiences and instrumental enrichment. *Special Services in the Schools, 3*(1-2), 49-82.

Feuerstein, R., Rand, Y., & Hoffman, M. B. (1979). *The dynamic assessment of retarded performers: The learning potential assessment device, theory, instruments, and techniques.* Glenview, IL: Scott, Foresman.

Feuerstein, R., Rand, Y., Hoffman, M. B., & Miller, R. (1980). *Instrumental enrichment: An intervention program for cognitive modifiability.* Baltimore, MD: University Park Press.

Feuerstein, R., Rand, Y., Jensen, M., Kaniel, S., Tzuriel, D., Ben Shachar, N., & Mintzker, Y. (1986a). Learning potential assessment. *Special Services in the Schools, 3*(1-2), 85-106.

Figueroa, R. A. (1990). Best practices in the assessment of bilingual children. In A. Thomas & J. Grimes (Eds.), *Best practices in school psychology II* (pp. 93-106). Washington, DC: National Association of School Psychologists.

Franzbach, M. (1965). *Lessings Huarte-Uebersetzung (1752): Die Rezeption und Wirkungsgeschichte des "Examen de Ingenios para las Ciencia" (1575) in Deutschland* [Lessing's translation (1752) of Huarte: History of the reception and impact of "Examen de Ingenios para las Ciencias" (1575) in Germany]. Hamburgz: Cram, de Gruyter.

Freire, P. (1970). *Pedagogy of the oppressed.* New York: Seabury.

French, J., & Hale, R. (1990). A history of the development of psychological and educational testing. In C. R. Reynolds & R. W. Kamphaus (Eds.), *Handbook of psychological and educational assessment of children's intelligence and achievement* (pp. 3-28). New York: Guilford.

Gaines, S., Jr., & Reed, E. (1995). Prejudice from Allport to DuBois. *American Psychologist, 50*(3), 103.

Galton, F. (1869). *Hereditary genius: An enquiry into its laws and consequences.* London: Collins.

Galton, F. (1883). *Inquiry into human faculty and its development.* London: Macmillan.

Gardner, H. (1983). *Frames of mind: The theory of multiple intelligences.* New York: Basic Books.

Gardner, H. (1989). Zero-based arts education: An introduction to Arts PROPEL. *Studies in Art Education, 30,* 71-83.

Gardner, H. (1993). *Multiple intelligences.* New York: Basic Books.

Gardner, H., & Hatch, T. (1989). Multiple intelligences go to school: Educational implications of the theory of multiple intelligences. *Educational Researcher, 18*(8), 4-10.

Gardner, H., Howard, V., & Perkins, D. (1974). Symbol systems: A philosophical, psychological and educational investigation. In D. Olson (Ed.), *Media and symbols* (pp. 37-55). Chicago: University of Chicago Press.

Gardner, H., & Wolf, D. (1983). Waves and streams of symbolization. In D. R. Rogers & J. A. Sloboda (Eds.), *The acquisition of symbolic skills* (pp. 19-42). London: Plenum.

Gauvain, M. (1995). Thinking in niches: Sociocultural influences on cognitive development. *Human Development, 38,* 25-45.

Gauvain, M., & Rogoff, B. (1989). Ways of speaking about space: The development of children's skill at communicating spatial knowledge. *Cognitive Development, 4,* 295-307.

Gay, J., & Cole, M. (1967). *The new mathematics and an old culture.* New York: Holt, Rinehart & Winston.

Geertz, C. (1973). *Interpretation of cultures.* New York: Basic Books.

Gladwin, H. (1971). *East is a big bird.* Cambridge, MA: Harvard University Press.

Glaser, R. (1977). *Adaptive education: Individual diversity and learning.* New York: Holt, Rinehart & Winston.

Glutting, J., & McDermott, P. (1990). Principles and problems in learning potential. In C. R. Reynolds & R. W. Kamphaus (Eds.), *Handbook of psychological and educational assessment of children's intelligence and achievement* (pp. 296-347). New York: Guilford.

Goodnow, J. J. (1976). The nature of intelligent behavior: Questions raised by cross-cultural studies. In L. B. Resnick (Ed.), *The nature of intelligence.* Hillsdale, NJ: Lawrence Erlbaum.

Goodnow, J. J. (1990). The socialization of cognition: What's involved? In J. W. Stigler, R. A. Shweder, & G. Herdt (Eds.), *Cultural psychology* (pp. 259-286). Cambridge, UK: Cambridge University Press.

Gopaul-McNicol, S. (1992a). Understanding and meeting the psychological and educational needs of African American and Spanish speaking students. *School Psychology Review, 21*(4), 529-531.

Gopaul-McNicol, S. (1992b). Implications for school psychologists: Synthesis of the miniseries. *School Psychology Review, 21*(4), 597-600.

Gopaul-McNicol, S. (1993). *Working with West Indian families.* New York: Guilford.]

Gopaul-McNicol, S., & Armour-Thomas, E. (1996, February). *A practical guide for enhancing the intellectual potential of children: A biocultural perspective.* Presented at the annual professional development workshops for school psychologists: New York City Board of Education: Brooklyn.

Gopaul-McNicol, S., & Armour-Thomas, E. (1997a). A bioecological case study: A Caribbean child. *Cultural Diversity and Mental Health, 3*(2), 145-151.

Gopaul-McNicol, S., & Armour-Thomas, E. (1997b). The role of bioecological assessment system in writing a culturally sensitive report: The importance of assessing other intelligences. *Journal of Social Distress and the Homeless, 6*(2), 129-141.

Gopaul-McNicol, S., Elizalde-Utnick, G., Nahari, S., & Louden, D. (1998). *A test review guide for bilingual children: Cognitive assessment.* New York: National Nursing League.

Gordon, E. W. (1977). Diverse human populations and problems in educational program evaluation via achievement testing. In M. J. Wargo & D. R. Green (Eds.), *Achievement testing of disadvantaged and minority students for educational program evaluation* (pp. 29-40). New York: CTB/McGraw-Hill.

Gordon, E. W. (1988). *Human diversity and pedagogy.* New Haven, CT: Yale University, Institute for Social and Policy Studies.

Gordon, E. W. (1991). Human diversity and pluralism. *Educational Psychologist, 26,* 99-108.

Gordon, E. W. (1995). Toward an equitable system of educational assessment. *Journal of Negro Education, 64*(3), 360-372.

Gordon, E. W., & Armour-Thomas, E. (1991). Culture and cognitive development. In L. Okagaki & R. J. Sternberg (Eds.), *Directors and development: Influences on the development of children's thinking.* Hillsdale, NJ: Lawrence Erlbaum.

Gordon, E. W., & Bonilla-Bowman, C. (1994). Equity and social justice in educational achievement. In R. Berne & L. O. Picus (Eds.), *Outcome equity in education.* Thousand Oaks, CA: Corwin Press.

Gordon, E. W., Miller, F., & Rollock, D. (1990). Coping with communicentric bias in knowledge production in the social sciences. *Educational Researcher, 19*(3), 14-19.

Gordon, E. W., & Shipman, S. (1979). Human diversity, pedagogy and educational equity. *American Psychologist, 34*(1), 1030-1036.

Gordon, E. W., & Terrell, M. (1981). The changed social context of testing. *American Psychologist, 36,* 1167-1171.

Gordon, R. A., & Rudert, E. E. (1979). Bad news concerning IQ tests. *Sociology of Education, 52,* 174-190.

Greenfield, P. M. (1974). Comparing dimensional categorization in natural and artificial contexts: A developmental study among the Zenacantecos of Mexico. *Journal of Social Psychology, 93,* 157-171.

Guberman, R., & Greenfield, P. M. (1991). Learning and transfer in everyday cognition. *Cognitive Development, 6,* 233-260.

Guilford, J. P. (1967). *The nature of human intelligence.* New York: McGraw-Hill.

Gustafsson, J.-E. (1984). A unifying model for the structure of intellectual abilities. *Intelligence, 8,* 179-203.

Guttierrez, J., & Sameroff, A. (1990). Determinants of complexity in Mexican-American and Anglo-American mothers' conceptions of child development. *Child Development, 61,* 384-394.

Hale, J. (1982). *Black children: Their roots, culture, and learning styles.* Provo, UT: Brigham Young University Press.

Hale-Benson, J. E. (1986). *Black children: Their roots, culture and learning styles* (Rev. ed.). Baltimore, MD: Johns Hopkins University Press.

Hamayan, E. V., & Damico, J. S. (Eds.). (1991). *Limiting bias in the assessment of bilingual students.* Austin, TX: Pro-Ed.

Harrison, A., Wilson, M., Pine, C., Chan, S., & Buriel, R. (1990). Family ecologies of ethnic minority children. *Child Development, 61,* 347-362.

Haynes, N. (1995). How skewed is the bell curve? *Journal of Black Psychology, 21*(3), 275-299.

Haynes, N., & Comer, J. (1993). The Yale School Development Program: Process, outcomes and policy implications. *Urban Education, 28*(2), 166-169.

Heath, S. B. (1983). *Ways with words: Language, life and work in communities and classrooms.* Cambridge, UK: Cambridge University Press.

Helms, J. E. (1989). Eurocentrism strikes in strange places and in unusual ways. *The Counseling Psychologist, 17,* 643-647.

Helms, J. E. (1992). Why is there no study of cultural equivalence in standardized cognitive ability testing? *American Psychologist, 47*(9), 1083-1101.

Herrnstein, R., & Murray, C. (1994). *The bell curve.* New York: Free Press.

Herrnstein, R., Nickerson, R., de Sanchez, M., & Swets, J. (1986). Teaching thinking skills. *American Psychologist, 41*(11), 1279-1289.

Hilliard, A. (1991). Do we have the will to educate all children? *Educational Leadership, 49*(1), 31-36.

Hilliard, A. (1996). Either a paradigm shift or no mental measurement: The nonscience and the nonsense of the bell curve. *Cultural Diversity and Mental Health, 2*(1), 1-20.

Hilliard, A. G., III. (1976). *Alternatives to IQ testing: An approach to the identification of "gifted" minority children.* Final report to the California State Department of Education, Special Education Support Unit. (ERIC Clearinghouse on Early Childhood Education No. ED 146 009)

Hilliard, A. G. (1979). Standardization and cultural bias as impediments to the scientific study and validation of "intelligence." *Journal of Research and Development in Education, 12*(2), 47-58.

Horn, J. L. (1965). *Fluid and crystallized intelligence: A factor analytic study of the structure among primary mental abilities.* Unpublished doctoral dissertation, University of Illinois. (University Microfilms No. 65-7113)

Horn, J. L. (1991a). Measurement of intellectual capabilities: A review of theory. In K. S. McGrew, J. K. Werder, & R. W. Woodcock (Eds.), *WJ-R technical manual.* Chicago: Riverside.

Horn, J. L. (1991b). Measurement of intellectual capabilities: A review of theory. In K. S. McGrew, J. K. Werder, & R. W. Woodcock (Eds.), *A reference on theory and current research to supplement the Woodcock-Johnson-Revised Examiner's Manuals* (pp. 197-245). Allen, TX: DLM.

Horowitz, F. D., & O' Brien, M. (1989). A reflective essay on the state of our knowledge and the challenges before us. *American Psychologist, 44,* 441-445.

Howe, K. R. (1992). Liberal democracy, equal educational opportunity and the challenge of multiculturalism. *American Educational Research Journal, 29*(3), 455-470.

Hunt, E. B. (1978). Mechanics of verbal ability. *Psychological Review, 85,* 109-130.

Intelligence and its measurement: A symposium. (1921). *Journal of Educational Psychology, 12,* 123-147, 195-216, 271-275.

Jensen, A. R. (1969). How much can we boost IQ and scholastic achievement? *Harvard Educational Review, 39*(1), 1-123.

Jensen, A. R. (1979). Outmoded theory or unconquered frontier? *Creative Science and Technology, 2,* 16-29.

Jensen, A. R. (1980). *Bias in mental testing.* New York: Free Press.

Jensen, A. R. (1987). Unconfounding genetic and nonshared environmental effects. *Behavioral and Brain Sciences, 10,* 26-27.

Jensen, A. R. (1991). General mental ability: From psychometrics to biology. *Diagnostique, 16,* 134-144.

Jensen, A. R., & Whang, P. A. (1994). Speed of accessing arithmetic facts in long term memory: A comparison of Chinese-American and Anglo-American Children. *Contemporary Educational Psychology, 19,* 1-12.

Johnson, S. D. (1982). *The Minnesota, multiethnic counselor education curriculum: The design and evaluation of an intervention for cross-cultural counselor education.* Unpublished doctoral dissertation, University of Minnesota, Minneapolis.

Johnson, S. D. (1987). Knowing that versus knowing how: Toward achieving expertise through multicultural training for counseling. *The Counseling Psychologist, 15,* 320-331.

Johnson, S. D. (1990). Towards clarifying culture, race and ethnicity in the context of multicultural counseling. *Journal of Multicultural Counseling and Development, 18*(4), 16-31.

Jones, L. V. (1985). Interpreting Spearman's general factor. *Behavioral and Brain Science, 8*(2), 233.

Kamphaus, R. (1993). *Clinical assessment of children's intelligence.* Boston: Allyn & Bacon.

Kaufman, A. S., & Kaufman, N. L. (1983). *K-ABC: Kaufman Assessment Battery for Children.* Circle Pines, MN: American Guidance Service.

Kearins, J. (1981). Visual spatial memory in Australian Aboriginal children of desert regions. *Cognitive Psychology, 13,* 434-460.

Keil, F. C. (1981). Constraints on knowledge and cognitive development. *Psychological Review, 88,* 197-227.

Keil, F. C. (1984). Mechanisms of cognitive development and the structure of knowledge. In R. J. Sternberg (Ed.), *Mechanisms of cognitive development*. San Francisco: Freeman.

Kiselica, M. S. (1991, September/October). Reflections on a multicultural internship experience. *Journal of Counseling and Development, 70,* 126-130.

Kornhaber, M., Krechevsky, M., & Gardner, H. (1990). Engaging intelligence. *Educational Psychologist, 25*(3/4), 177-199.

Krechevsky, M., & Gardner, H. (1990). The emergence and nurturance of multiple intelligences: The Project Spectrum Approach. In M. J. A. Howe (Ed.), *Encouraging the development of exceptional skills and talents* (pp. 222-245). Leicester, UK: British Psychological Society.

Laboratory of Comparative Human Cognition. (1982). Culture and intelligence. In R. J. Sternberg (Ed.), *Handbook of human intelligence* (pp. 642-719). New York: Cambridge University Press.

Lancy, D. F., & Strathern, A. J. (1981). Making two's: Pairing as an alternative to the taxonomic mode of representation. *American Anthropologist, 83,* 773-795.

Lantz, D. A. (1979). A cross-cultural comparison of communication abilities: Some effects of age, schooling and culture. *International Journal of Psychology, 14,* 171-183.

Laosa, L. M. (1980). Maternal teaching strategies in Chicano and Anglo-American families: The influence of culture and education on maternal behavior. *Child Development, 51,* 759-765.

Laosa, L. M. (1981). Maternal behavior: Sociocultural diversity in modes of family interaction. In R. W. Henderson (Ed.), *Parent-child interaction: Theory, research, and prospects* (pp. 12-167). New York: Academic Press.

Lave, J. (1977). Tailor-made experiments and evaluating the intellectual consequences of apprenticeship training. *Quarterly Newsletter of the Institute for Comparative Human Development, 1,* 1-3.

Lave, J. (1988). *Cognition in practice: Mind, mathematics and culture in everyday life.* Cambridge, UK: Cambridge University Press.

Lave, J., Murtaugh, M., & de la Roche, D. (1984). The dialectic of arithmetic in grocery shopping. In B. Rogoff & J. Lave (Eds.), *Everyday cognition: Its development in social context.* Cambridge, MA: Harvard University Press.

Lave, J., & Wenger, E. (1991). *Situated learning: Legitimate peripheral participation.* Cambridge, UK: Cambridge University Press.

Lee Katz, L. (1991). Cultural scripts: The home-school connection. *Early Child Development and Care, 73,* 95-102.

Lewin, K. (1935). *A dynamic theory of personality.* New York: McGraw-Hill.

Lidz, C. S. (Ed.). (1987). *Dynamic assessment.* New York: Guilford.

Lidz, C. S. (1991). *Practitioner's guide to dynamic assessment.* New York: Guilford.

Linden, K. W., & Linden, J. D. (1968). *Modern mental measurement: A historical perspective.* Boston: Houghton Mifflin.

Lipsky, D. R., & Gartner, A. (1996). Inclusion, school restructuring, and the remaking of American society. *Harvard Educational Review, 66*(4), 762-796.

Lonner, W. J. (1981). Psychological tests and intercultural counseling. In P. B. Pedersen, J. G. Draguns, W. J. Lonner, & J. E. Trimbie (Eds.), *Counseling across cultures* (pp. 275-303). Honolulu: East West Center/University of Hawaii.

Mackie, D. (1980). A cross-cultural study of intra- and interindividual conflicts of concentrations. *European Journal of Social Psychology, 10,* 313-318.

Mackie, D. (1983). The effect of social interaction on conservation of spatial relations. *Journal of Cross-Cultural Psychology, 14,* 131-151.

Maker, C. J. (1992, Fall). Intelligence and creativity in multiple intelligences: Identification and development. *Educating Able Learners,* pp. 12-19.

Matthews, J. (1989, March 25). Aspiring lawyers already finding a way to mak a point. *The Washington Post,* p. A3.

Mbiti, J. (1970). *African religions and philosophy.* Garden City, NJ: Anchor.

McGrew, K., & Flanagan, D. (1995). *An intelligence test desk reference: The cross-battery approach to test interpretation.* Paper presented at the National Association of School Psychologist Convention, Atlanta.

McGrew, K. S. (1994). *Clinical interpretation of the Woodcock Johnson Tests of Cognitive Ability—Revised.* Boston: Allyn & Bacon.

McGrew, K. S. (1995). Analysis of the major intelligence batteries according to a proposed comprehensive Gf-Gc framework of human cognitive and knowledge abilities. In D. P. Flanagan, J. L. Genshaft, & P. L. Harrison (Eds.), *Beyond traditional intellectual assessment: Contemporary and merging theories, tests and issues.* New York: Guilford.

McLoyd, V. (1990). Minority children: Introduction to the special issue. *Child Development, 61,* 263-266.

Mercer, J. R. (1979). In defense of racially and culturally nondiscriminatory assessment. *School Psychology Digest, 8*(1), 89-115.

Messick, S. (1976). Personality consistencies in cognition and creativity. In S. Messick (Ed.), *Individuality in learning* (pp. 4-22). San Francisco: Jossey-Bass.

Messick, S., & Anderson, S. (1970). Educational testing individual development and social responsibility. *The Counseling Psychologist,* 2(2), 93-97.

Miller-Jones, D. (1989). Culture and testing. *American Psychologist, 44,* 360-366.

Missiuna, C., & Samuels, M. (1988). Dynamic assessment: Review and critique. *Special Services in the Schools, 5*(1-2), 1-22.

Moon, J. D. (1993). *Constricting community: Moral pluralism and tragic conflicts.* Princeton, NJ: Princeton University Press.

Moyer, J. (1986, May/June). Child development as a base for decision making. *Childhood Education,* pp. 325-329.

Munroe, R. H., Munroe, R. L., & Whiting, B. B. (Eds.). (1981). *Handbook of cross-cultural human development.* New York: Garland.

Munroe, R. L., & Munroe, R. H. (1971). Effect of environmental experience on spatial ability in an East African society. *Journal of Social Psychology, 125*, 23-33.

Murray, B. (1996). Developing phoneme awareness through books. *Reading and Writing: An Interdisciplinary Journal, 8*(4), 307-322.

Murray, H. A. (1938). *Explorations in personality.* New York: Oxford University Press.

Murrone, J., & Gynther, M. (1991). Teachers' implicit "theories" of children's intelligence. *Psychological Reports, 69*, 1195-1201.

Murtaugh, M. (1985, Fall). The practice of arithmetic by American grocery shoppers. *Anthropology and Education Quarterly*, 23.

Niesser, U. (1976). Genera, academic, and artificial intelligence. In L. Resnick (Ed.), *Human intelligence: Perspectives on its theory and measurement* (pp. 179-189). Norwood, NJ: Ablex.

Neisser, U. (1979). The concept of intelligence. *Intelligence, 3*, 217-227.

Neisser, U., Boodoo, G., Bouchard, T., Boykin, W., Brody, N., Ceci, S., Halpern, D. F., Loehlin, J. C., Perloff, R., Sternberg, R., & Urbina, S. (1996). Intelligences: Knowns and unknowns. *American Psychologist, 51*(2), 77-101.

Nerlove, S. B., & Snipper, A. S. (1981). Cognitive consequences of cultural opportunity. In R. H. Munroe, R. L. Munroe, & B. B. Whiting (Eds.), *Handbook of cross-cultural human development.* New York: Garland.

Nettelbeck, T. (1985). What reaction times time. *Behavioral and Brain Science, 8*(2), 235-236.

Newman, F., & Holzman, L. (1993). *Lev Vygotsky: Revolutionary scientist.* London: Routledge.

Nichols, R. (1981). Origins, nature and determinants of intellectual development. In M. Begab, H. C. Haywood, & H. Garber (Eds.), *Psychosocial determinants of retarded performance, Vol. 1.* Baltimore, MD: University Park Press.

Noble, C. E. (1969). Race, reality and experimental psychology. *Perspectives in Biology and Medicine, 13*, 10-30.

Nobles, W. (1980). African philosophy: Foundations for black psychology. In R. Jones (Ed.), *Black psychology* (pp. 23-35). New York: Harper & Row.

Oakes, J. (1990). *Multiplying inequalities: The effects of race, social class, and tracking on opportunities to learn mathematics and science.* Santa Monica: Rand Corporation.

Obringer, S. J. (1988, November). *A survey of perceptions by school psychologists of the Stanford-Binet IV.* Paper presented at the meeting of the Mid-South Educational Research Association, Louisville, KY.

Ochs, E., & Schiefflin, B. (1984). Language acquisition and socialization: Three developmental stories and their implications. In R. Shweder & R. LeVine (Eds.), *Culture and its acquisition.* Chicago: University of Chicago Press.

Ogbu, J. U. (1986). The consequences of the American caste system. In *The school achievement of minority children: New perspectives.* London: Lawrence Erlbaum.

Ogbu, J. U. (1987). Variability in minority responses to schooling: Nonimmigrants vs. immigrants. In G. Spindler (Ed.), *Interpretive ethnography of education at home and abroad* (pp. 255-278). Hillsdale, NJ: Lawrence Erlbaum.

Okagaki, L., & Sternberg, R. J. (1993). Parental beliefs and children's school performance. *Child Development, 64,* 36-56.

Palinscar, A. S., & Brown, A. L. (1984). Reciprocal teaching of comprehension-fostering and comprehension-monitoring activities. *Cognition and Instruction, 1,* 117-175.

Parke, R. D., & Bhavnagri, N. P. (1989). Parents as managers of children's peer relationships. In D. Belle (Ed.), *Children's social networks and social supports.* New York: John Wiley.

Parker, W. M., Valley, M. M., & Geary, C. A. (1986). Acquiring cultural knowledge for counselors in training: A multifaceted approach. *Counselor Education and Supervision, 26,* 61-71.

Pellegrino, J. W., & Glaser, R. (1979). Cognitive correlates and components in the analysis of individual differences. In R. J. Sternberg & D. K. Detterman (Eds.), *Human intelligence: Perspectives on its theory and its measurement.* Norwood, NJ: Ablex.

Persell, C. H. (1977). *Education and inequality: The roots and results of stratification in America's schools.* New York: Free Press.

Phinney, J. S. (1996, September). When we talk about American ethnic groups, what do we mean? *American Psychologist, 51*(9), 918-927.

Piaget, J. (1952). *The origins of intelligence in children.* New York: International University Press.

Plomin, R. (1985). Behavioral genetics. In D. Detterman (Ed.), *Current topics in human intelligence* (Vol. 1). Norwood, NJ: Ablex.

Poortinga, Y., van de Vijver, F., Joe, R., & van de Koppel, J. (1989). Peeling the onion called culture: A synopsis. In C. Kagitcibasi (Ed.), *Growth and progress in cross-cultural psychology* (pp. 22-34). Berwyn, PA: Swets North American.

Rawls, J. (1973). *A theory of justice.* London: Oxford University Press.

Rawls, J. (1993). *Political liberalism.* New York: Columbia.

Ribeiro, J. L. (1980). Testing Portuguese immigrant children: Cultural patterns and group differences in responses to the WISC-R. In D. P. Macedo (Ed.), *Issues in Portuguese bilingual education* (pp. 90-101). Cambridge, MA: National Assessment and Dissemination Center for Bilingual Education.

Ridley, C. R. (1985). Imperatives for ethnic and cultural relevance in psychology training programs. *Professional Psychology: Research and Practice, 16*(5), 611-622.

Riley, M. K., Morocco, C. C., Gordon, S. M., & Howard, C. (1993). Walking the talk: Putting constructivist thinking into practice of constructivist principles. *Educational Horizons, 71*(4), 187-196.

Rivers, W. H. R. (1926). *Psychology and ethnology.* New York: Harcourt Brace.

Robinson, D. (1994). Philosophical views of intelligence. In *Encyclopedia of human intelligence.* New York: Macmillan.

Rogoff, B. (1978). Spot observations: An introduction and examination. *Quarterly Newsletter of the Institute for Comparative Human Development, 2,* 21-26.

Rogoff, B. (1981a). Schooling's influence on memory test performance. *Child Development, 52,* 260-267.

Rogoff, B. (1981b). Schooling and the development of cognitive skills. In H. Triandis & A. Heron (Eds.), *Handbook of cross-cultural psychology, Vol. 4* (pp. 233-294). Rockleigh, NJ: Allyn & Bacon.

Rogoff, B. (1990). *Apprenticeship in thinking.* New York: Oxford University Press.

Rogoff, B., & Chavajay, P. (1995). What's become of research on the cultural basis of cognitive development? *American Psychologist, 50*(10), 859-877.

Rogoff, B., & Waddell, K. J. (1982). Memory for information organized in a scene by children from two cultures. *Child Development, 53,* 1224-1228.

Rosenbaum, J. E. (1980). Social implications of educational grouping. *Review of Research in Education, 8,* 361-401.

Rueda, R., & Matinez, I. (1992). Fiesta educativa: One community's approach to parent training in developmental disabilities for Latino families. *Journal of the Association of Severe Handicaps, 17*(2), 95-103.

Samuda, R. (1975). From ethnocentrism to a multicultural perspective in educational testing. *Journal of Afro-American Issues, 3*(1), 4-17.

Sattler, J. M. (1988). *Assessment of children* (3rd ed.). San Diego: Author.

Saxe, G. (1988). The mathematics of street vendors. *Child Development, 59,* 1415-1425.

Saxe, G. (1991). *Culture and cognitive development: Studies in mathematical understanding.* Hillsdale, NJ: Lawrence Erlbaum.

Scarr, S., & Carter-Salzman, L. (1982). Genetics and intelligence. In R. J. Sternberg (Ed.), *Handbook of human intelligence* (pp. 792-896). Cambridge, UK: Cambridge University Press.

Scarr, S., & Ricciuti, A. (1991). What effects do parents have on their children? In L. Okagaki & R. J. Sternberg (Eds.), *Directors of development: Influences on the development of children's thinking.* Hillsdale, NJ: Lawrence Erlbaum.

Schaefer, E. S. (1987). Parental modernity and child academic competence: Towards a theory of individual and societal development. *Early Development and Care, 27,* 373-389.

Separate and unequal. (1993, December 13). *U.S. News & World Report,* pp. 46-60.

Serpell, R. (1979). How specific are perceptual skills? A cross-cultural study of pattern reproduction. *British Journal of Psychology, 70,* 365-380.

Serpell, R., Baker, L., Sonnenschein, S., & Hill, S. (1993, May). *Contexts for the early appropriation of literacy: Caregiver meanings of recurrent activities.* Paper presented at the annual meeting of the American Psychological Society, Chicago.

Shade, B. J. (1982). Afro-American cognitive style: A variable in school success? *Review of Educational Research, 52*(2), 219-244.

Shearer, D., & Loftin, C. (1984). The Portage Project. In R. Dangel & R. Polster (Eds.), *Parent foundations of* (p. 93). New York: Guilford Press.

Shipman, S., & Shipman, S. (1985). Cognitive styles: Some conceptual, methodological and applied issues. In E. W. Gordon (Ed.), *Review of research in education* (Vol. 12, pp. 229-291). Washington, DC: American Educational Research Association.

Siegel, L. S. (1984). Home environmental influence on cognitive development in pre-term and full-term children during the first five years. In A. W. Gottfried (Ed.), *Home environment and early cognitive development* (pp. 19- 34). Orlando, FL: Academic Press.

Siegel, M. (1991). *Knowing children: Experiments in conversation and cognition.* London: Lawrence Erlbaum.

Skrtic, T. M. (1991). The special education paradox: Equity as the way to excellence. *Harvard Educational Review, 61*(2), 148-206.

Slavin, R. E. (1987). *A review of research on elementary ability grouping.* Baltimore, MD: Johns Hopkins University Press.

Spearman, C. (1923). *The nature of "intelligence" and the principles of cognition.* London: Macmillan.

Spearman, C. (1927). *The abilities of man: Their nature and measurement.* New York: Macmillan. (Reprinted 1981, New York: AMS)

Steinberg, L., Dornbusch, S. M., & Brown, B. B. (1992). Ethnic differences in adolescent achievement: An ecological perspective. *American Psychologist, 47*(6), 723-729.

Sternberg, R. J. (1977a). *Intelligence, information processing, and analogical reasoning: The componential analysis of human abilities.* Hillsdale, NJ: Lawrence Erlbaum.

Sternberg, R. J. (1977b). Component processes in analogical reasoning. *Psychological Review, 84,* 353-378.

Sternberg, R. J. (1980). Sketch of a componential subtheory of human intelligence. *Behavioral and Brain Sciences, 3,* 573-584.

Sternberg, R. J. (1984, January). What should intelligence tests test? Implications of a triarchic theory of intelligence for intelligence testing. *Educational Researcher,* pp. 5-15.

Sternberg, R. J. (1985a). Teaching critical thinking, Part 1: Are we making critical mistakes? *Phi Delta Kappan, 67*(3), 104-108.

Sternberg, R. J. (1985b). Teaching critical thinking, Part 2: Possible solutions. *Phi Delta Kappan, 67*(4), 277-280.

Sternberg, R. J. (1985c). *Beyond IQ: A triarchic theory of human intelligence.* New York: Cambridge University Press.

Sternberg, R. J. (1986). *Intelligences applied.* New York: Harcourt Brace Jovanovich.

Sternberg, R. J. (Ed.). (1988). *The nature of creativity: Contemporary psychological perspectives.* New York: Cambridge University Press.

Sternberg, R. J., Conway, B. E., Ketron, J. L., & Bernstein, M. (1981). People's conception of intelligence. *Journal of Personality and Social Psychology, 41,* 37-55.

Sternberg, R. J., & Davidson, J. (1989). A four-prong model for intellectual development. *Journal of Research and Development in Education, 22*(3), 22-28.

Sternberg, R. J., & Detterman, D. K. (Eds.). (1986). *What is intelligence? Contemporary viewpoints on its nature and definition.* Norwood, NJ: Ablex.

Sternberg, R. J., & Powell, J. S. (1983). Comprehending verbal comprehension. *American Psychologist, 38,* 878-893.

Sternberg, R. J., Powell, J. S., & Kaye, D. B. (1982). The nature of verbal comprehending. *Poetics, 11,* 155-187.

Sternberg, R. J., & Wagner, R. K. (1986). *Practical intelligence: Nature and origin of competence in the everyday world.* New York: Cambridge University Press.

Sternberg, R. J., Wagner, R. K., & Okagaki, L. (1993). Practical intelligence: The nature and role of tacit knowledge in work and at school. In H. Reese & J. Puckett (Eds.), *Advances in lifespan development* (pp. 205-227). Hillsdale, NJ: Lawrence Earlbaum.

Sternberg, R. J., Wagner, R. K., Williams, W. M., & Horvath, J. A. (1995). Testing common sense. *American Psychologist, 50*(11), 912-926.

Stigler, J. W. (1984). "Mental abacus": The effect of abacus training on Chinese children's mental calculation. *Cognitive Psychology, 16,* 145-176.

Stocking, G. (1968). *Race, culture and evolution.* New York: Free Press.

Strom, R., Johnson, A., Strom, S., & Strom, P. (1992a). Designing curriculum for parents of gifted children. *Journal for the Education of the Gifted, 15*(2), 182-200.

Strom, R., Johnson, A., Strom, S., & Strom, P. (1992b). Educating gifted Hispanic children and their parents. *Hispanic Journal of Behavioral Sciences, 14*(3), 383-393.

Sue, S. (1991). Ethnicity and culture in psychological research and practice. In J. Goodchilds (Ed.), *Psychological perspectives on human diversity in America* (pp. 51-85). Washington, DC: American Psychological Association.

Super, C. M. (1980). Cognitive development: Looking across at growing up. *New Directions for Child Development: Anthropological Perspectives on Child Development, 8,* 59-69.

Super, C. M., & Harkness, S. (1986). The development niche: A conceptualization at the interface of child and culture. *International Journal of Behavioral Development, 9,* 545-569.

Surber, J. (1995). Best practices in a problem-solving approach to psychological report writing. In Thomas, A., & Grimes, J. (Eds.). (1977). *Best practices in school psychology III.* Washington, DC: The National Association of School Psychologists.

Tallent, N. (1993). *Psychological report writing* (4th ed.). Englewood Cliffs, NJ: Prentice Hall.

Taylor, C. W. (1988). Various approaches to the definitions of creativity. In R. J. Sternberg (Ed.), *The nature of creativity: Contemporary psychological perspectives* (pp. 37-49). New York: Cambridge University Press.

Thomas, A., & Chess, S. (1977). *Temperament and development.* New York: Brunner/Mazel.

Thomas, T. (1990, August). *Is it learning disability, mental retardation, or educational deprivation: An exploration with Hispanic children?* Paper presented at the American Psychological Association Annual Convention, Boston.

Thompson, R. W., & Hixson, P. (1984). Teaching parents to encourage independent problem solving in preschool-age children. *Language, Speech and Hearing Services in the Schools, 15,* 175-181.

Thorndike, R. L., Hagen, E. P., & Sattler, J. M. (1986). *Stanford-Binet Intelligence Scale: Fourth edition.* Chicago: Riverside.

Thurstone, L. L. (1924). *The nature of intelligence.* New York: Harcourt Brace.

Thurstone, L. L. (1938). Primary mental abilities. *Psychometrika Monographs, 1.*

Tuck, K. (1985). *Verve inducement effects: The relationship of task performance to stimulus variability and preference in working class black and white schoolchildren.* Unpublished Doctoral Dissertation, Howard University, Washington, DC.

Tylor, E. B. (1874). *Primitive culture.* London: John Murray.

Uttal, D. H., & Wellman, H. M. (1989). Young children's representation of spatial information acquired from maps. *Developmental Psychology, 25,* 128-138.

Valencia, R., Henderson, R., & Rankin, R. (1981). Relationship of family constellation and schooling to intellectual performance of Mexican American children. *Journal of Educational Psychology, 73*(4), 524-532.

Van Daalen-Kapteijns, M. M., & Elshout-Mohr, M. (1981). The acquisition of word meaning as a cognitive learning process. *Journal of Verbal Learning and Verbal Behavior, 20,* 386-399.

Vaughn, B. E., Block, J. E., & Block, J. (1988). Parental agreement on child-rearing during early childhood and the psychological characteristics of adolescents. *Child Development, 59,* 1020-1033.

Vernon, P. A. (1990). The use of biological measures to estimate behavioral intelligence. *Educational Psychologist, 25*(3-4), 293-304.

Vygotsky, L. S. (1978). *Mind in society: The development of higher psychological processes.* Cambridge, MA: Harvard University Press.

Wechsler, D. (1944). *The measurement of adult intelligence* (3rd ed.). Baltimore, MD: Williams & Wilkins.

Wechsler, D. (1958). *The measurement and appraisal of adult intelligence* (4th ed.). Baltimore, MD: Williams & Wilkins.

Wechsler, D. (1974). *Manual for the Wechsler Intelligence Scale for Children—Revised (WISC-R).* New York: Psychological Corporation.

Weinberg, R. A. (1989). Intelligence and IQ: Landmark issues and great debates. *American Psychologist, 44*(2), 98-104.

Wertsch, J. V. (1979). From social interaction to higher psychological processes: A clarification and application of Vygotsky's theory. *Human Development, 22,* 1-22.

Wertsch, J. V. (1985). *Culture, communication, and cognition: Vygotskian perspectives.* Cambridge, UK: Cambridge University Press.

Whimbey, A. (1975). *Intelligence can be taught.* New York: Dutton.

Whimbey, A., & Lochhead, J. (1982). *Problem solving and comprehension: How to sharpen your thinking skills and increase your IQ.* Philadelphia: Franklin Institute.

Whitehurst, G., Fischel, J., Lonigan, C., Valdez-Menchaca, M., Arnold, D., & Smith, M. (1991). Treatment of early expressive language delay: If, when and how. *Topics in Language Disorders, 11*(4), 55-68.

Whiting, B. (1976). The problem of the packaged variable. In K. Riegel & J. Meacham (Eds.), *The developing individual in a changing world* (pp. 303-309). Chicago, Aldine.

Whiting, B. (1980). Culture and social behavior: A model for the development of social behavior. *Ethos, 8,* 95-116.

Williams, R. (1970). Danger: Testing and dehumanizing black children. *Clinical Child Psychology Newsletter, 9*(1), 5-6.

Williams, R. (1971, Spring). Abuses and misuses in testing black children. *Washington University Magazine, 41*(3), 34-37.

Williams, R. (1975). The BITCH-100: A culture-specific test. *Journal of Afro-American Issues, 3,* 103-106.

Witt, J. C., & Gresham, F. M. (1985). Review of Wechsler Intelligence Scale of Children-Revised. In J. V. Mitchell (Ed.), *The ninth mental measurement yearbook* (Vol. 2, pp. 1715-1716). Lincoln, NE: Buros Institute of Mental Measurements.

Wober, M. (1972). Culture and the concept of intelligence: A case in Uganda. *Journal of Cross-Cultural Psychology, 3,* 327-328.

Wober, M. (1974). Toward an understanding of the Kiganda concept of intelligence. In J. W. Berry & P. R. Dasen (Eds.), *Culture and cognition: Readings in cross-cultural psychology.* London: Methuen.

Woodcock, R. W. (1990). Theoretical foundations of the WJ-R measures of cognitive ability. *Journal of Psychoeducational Assessment, 8,* 231-258.

Woodcock, R. W., & Johnson, M. B. (1977). *Woodcock-Johnson Psycho-Educational Battery.* Chicago: Riverside.

Woodcock, R. W., & Johnson, M. B. (1989). *Woodcock-Johnson Psycho-Educational Battery-Revised.* Chicago: Riverside.

Zajonc, R. (1976). Family configuration and intelligence. *Science, 192,* 227-236.

Zuckerman, M. (1990). Some dubious premises in research and theory on racial differences: Scientific, social, and ethical issues. *American Psychologist, 45,* 1297-1303.

Name Index

Subject Index

About the Authors

Eleanor Armour-Thomas, originally from Trinidad and Tobago, West Indies, is Associate Professor in the School of Education at Queens College, City University of New York. She graduated from the University of the West Indies. From 1978 to 1984, she attended Columbia University, where she received a master's degree in applied human development and guidance, a master's degree in education, and a master's degree in behavioral analysis. In 1984, she received her doctorate in education psychology (schooling). Following completion of her graduate work, she continued her postdoctoral studies at Yale University. Her publications are in the areas of intellectual assessment of children from culturally diverse backgrounds and assessment of teaching in mathematics. In addition, as a consultant, she does evaluative research for educators and policymakers in several states.

Sharon-ann Gopaul-McNicol, originally from Trinidad and Tobago, West Indies, is Associate Professor in the School of Education at Howard University. She is an international expert in multicultural assessment. During the past 10 years, she assessed children both nationally and internationally. She is the author of three cross-cultural books and several journal articles, and she has given presentations throughout the world on electronic media. She has a bachelor's degree in psychology from New York University (1981), a master's degree in child and adolescent psychology from Columbia University (1982), a master's degree in general psychology from Hofstra University (1984), and a master's degree in school psychology from Hofstra University (1985). In 1986, she completed her doctorate in clinical psychology at Hofstra University.